Mastering Essential Math Skills

Geometry

Richard W. Fisher

★★★★★ Free Gift for a 5-Star Amazon Review!
We hope you enjoy this book. We love 5-star Amazon reviews. If you love this book and give it a 5-star review, as a token of our appreciation we would like to present you with a free gift. Just email us at **math.essentials@verizon.net** and let us know the name you used to post the review, and which book you reviewed, and we here at Math Essentials will send you a free gift (an $18.95 value).

Tell your friends about our award-winning math materials, like us on Facebook and follow us on Twitter. You are our best advertising, and we appreciate you!

Notes to the Teacher or Parent

What sets this book apart from other books is its approach. It is not just a math book, but a system of teaching math. Each daily lesson contains three key parts: **Review Exercises**, **Helpful Hints**, and **Problem Solving**. Teachers have flexibility in introducing new topics, but the book provides them with the necessary structure and guidance. The teacher can rest assured that essential math skills in this book are being systematically learned.

This easy-to-follow program requires only fifteen or twenty minutes of instruction per day. Each lesson is concise and self-contained. The daily exercises help students to not only master math skills, but also maintain and reinforce those skills through consistent review - something that is missing in most math programs. Skills learned in this book apply to all areas of the curriculum, and consistent review is built into each daily lesson. Teachers and parents will also be pleased to note that the lessons are quite easy to correct.

This book is based on a system of teaching that was developed by a math instructor over a thirty-year period. This system has produced dramatic results for students. The program quickly motivates students and creates confidence and excitement that leads naturally to success.

Please read the following "How to Use This Book" section and let this program help you to produce dramatic results with children and math students.

How to Use This Book

This book is best used on a daily basis. The first lesson should be carefully gone over with students to introduce them to the program and familiarize them with the format. It is hoped that the program will help your students to develop an enthusiasm and passion for math that will stay with them throughout their education.

As you go through these lessons every day, you will soon begin to see growth in the student's confidence, enthusiasm, and skill level. The students will maintain their mastery through the daily review.

Step 1

The students are to complete the review exercises, showing all their work. After completing the problems, it is important for the teacher or parent to go over this section with the students to ensure understanding.

Step 2

Next comes the new material. Use the "Helpful Hints" section to help introduce the new material. Be sure to point out that it is often helpful to come back to this section as the students work independently. This section often has examples that are very helpful to the students.

Step 3

It is highly important for the teacher to work through the two sample problems with the students before they begin to work independently. Working these problems together will ensure that the students understand the topic, and prevent a lot of unnecessary frustration. The two sample problems will get the students off to a good start and will instill confidence as the students begin to work independently.

Step 4

Solutions are located in the back of the book. Teachers may correct the exercises if they wish, or have the students correct the work themselves.

Table of Contents

Review Exercises

Note to students and teachers: This section will include daily review from all topics covered in this book. Here are some simple problems with which to get started.

1. 345
     16
  + 724
   1085

2. 715
  − 79
     636

3. 247
   ×   6
    1482

4. $96 + 72 + 16 =$

5. $800 - 216 =$

6. $8 × 394 =$

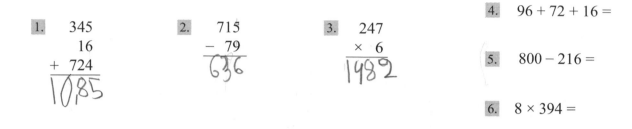

Helpful Hints	*Geometric Term:*	**Point**	**Line**	**Plane**	**Line Segment**	**Ray**
	Example:	• P	A B		A B	A B
	Symbol:	P	$\overleftrightarrow{AB}$	plane ABC	$\overline{AB}$	$\overrightarrow{AB}$

Use the figure to answer the following:

S1. Name 4 points

S2. Name 5 line segments

1. Name 5 lines

2. Name 5 rays

3. Name 3 points on $\overleftrightarrow{FD}$

4. Give another name for $\overleftrightarrow{AB}$

5. Give another name for $\overleftrightarrow{ED}$

6. Give another name for $\overleftrightarrow{AC}$

7. Name 2 line segments on $\overleftrightarrow{FD}$

8. Name 2 rays on $\overleftrightarrow{FE}$

9. Name 2 rays on $\overleftrightarrow{AC}$

10. What point is common to lines $\overleftrightarrow{FD}$ and $\overleftrightarrow{BE}$?

1.

2.

3.

4.

5.

6.

7.

8.

9.

10.

Score

Problem Solving

Ken earned 2,500 dollars in March and 3,752 dollars in April.
What were his total earnings for the two months?

Review Exercises

1. $724 + 16 + 347 =$

2. $506 - 397 =$

3. $7 \times 2,137 =$

4. $\begin{array}{r} 46 \\ \times \ 23 \\ \hline \end{array}$

5. $\begin{array}{r} 753 \\ 66 \\ 124 \\ + \ 237 \\ \hline \end{array}$

6. $\begin{array}{r} 5,000 \\ - \ 787 \\ \hline \end{array}$

Use what you have learned to answer the following questions.

Use the figure to answer the following:

S1. Name 4 points

S2. Name 5 lines

1. Name 5 line segments

2. Name 5 rays

3. Name 3 points on $\overleftrightarrow{GC}$

4. Give another name for $\overleftrightarrow{HD}$

5. Give another name for $\overleftrightarrow{JI}$

6. Give another name for $\overrightarrow{ED}$

7. Name 2 rays on $\overleftrightarrow{FB}$

8. Name 2 line segments on $\overleftrightarrow{GC}$

9. Name 2 rays on $\overleftrightarrow{GC}$

10. What point is common to $\overleftrightarrow{ID}$ and $\overleftrightarrow{AE}$?

1.

2.

3.

4.

5.

6.

7.

8.

9.

10.

Score

Problem Solving

A factory can produce 350 cars per week. How many cars can the factory produce in one year? (Hint: How many weeks are there in a year?)

Review Exercises

1. $3\overline{)636}$

2. $5\overline{)617}$

3. $8\overline{)2,372}$

4. $7 \times 658 =$

5. $926 + 75 + 396 =$

6. $7,001 - 2,658 =$

Helpful Hints

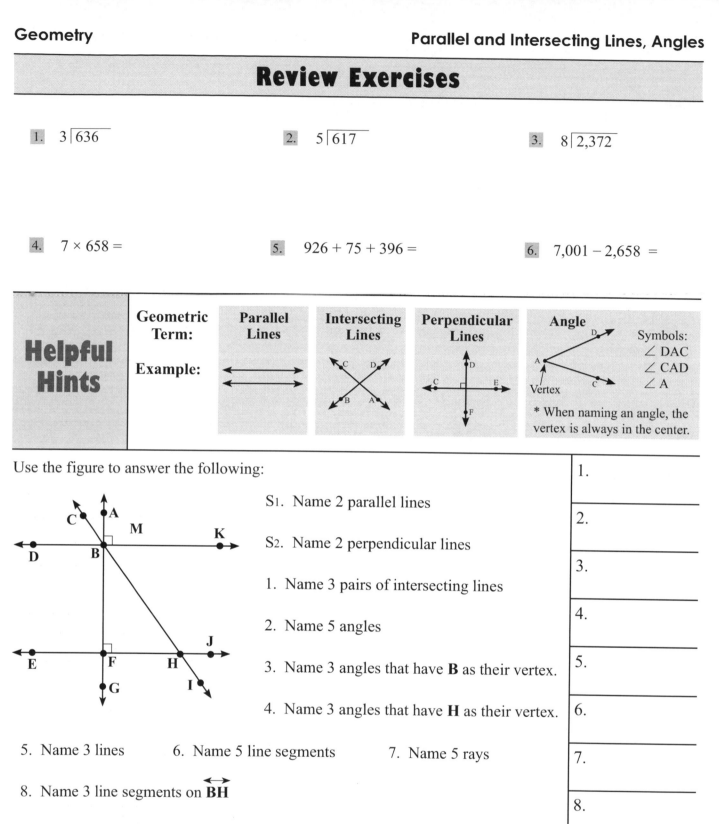

Geometric Term:	Parallel Lines	Intersecting Lines	Perpendicular Lines	Angle

Example:

Angle — Symbols: ∠ DAC, ∠ CAD, ∠ A — Vertex

* When naming an angle, the vertex is always in the center.

Use the figure to answer the following:

S1. Name 2 parallel lines

S2. Name 2 perpendicular lines

1. Name 3 pairs of intersecting lines

2. Name 5 angles

3. Name 3 angles that have **B** as their vertex.

4. Name 3 angles that have **H** as their vertex.

5. Name 3 lines   6. Name 5 line segments   7. Name 5 rays

8. Name 3 line segments on $\overleftrightarrow{BH}$

9. Name 3 lines which include point **B**.

10. Give two other names for ∠**JHI**

1.
2.
3.
4.
5.
6.
7.
8.
9.
10.

Problem Solving

Julio had test scores of 75, 96, 83, and 94.
What was his average score?

Score

Review Exercises

1. Sketch two parallel lines.

2. Sketch an angle and label it $\angle ABC$.

3. Sketch 2 lines $\overleftrightarrow{AB}$ and $\overleftrightarrow{CD}$ that are perpendicular.

4.
$$\begin{array}{r} 906 \\ \times\ \ 8 \\ \hline \end{array}$$

5.
$$\begin{array}{r} 7,112 \\ -\ \ 667 \\ \hline \end{array}$$

6. $7\overline{)847}$

Helpful Hints

* When identifying an angle, the vertex is always in the center.

Example: In $\angle CKD$, K is the vertex.

Use the figure to answer the following:

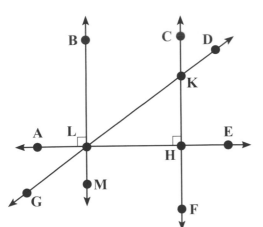

S1. Name 2 perpendicular lines.

S2. Name 2 parallel lines.

1. Name 4 angles.

2. Name 3 pairs of intersecting lines.

3. Name 2 angles with **L** as a vertex.

4. Name 2 angles with **K** as a vertex.

5. Name 4 lines.

6. Name 4 rays.

7. Name 4 line segments.

8. Name 3 lines which include point **H**.

9. Name 3 line segments on $\overleftrightarrow{AE}$.

10. Give another name for angle $\angle$**BLH**.

1.
2.
3.
4.
5.
6.
7.
8.
9.
10.
Score

Problem Solving

Six students earned \$924. If they wanted to divide the money equally, how much would each of them receive?

Review Exercises

1. $852 + 276 + 19 =$

2. Sketch 2 lines $\overleftrightarrow{AB}$ and $\overleftrightarrow{CD}$ that are parallel.

3. $800 - 65 =$

4. Sketch 2 lines $\overleftrightarrow{MK}$ and $\overleftrightarrow{CD}$ that are intersecting.

5.
$$\begin{array}{r} 65 \\ \times\ 52 \\ \hline \end{array}$$

6.
$$\begin{array}{r} 246 \\ \times\ 70 \\ \hline \end{array}$$

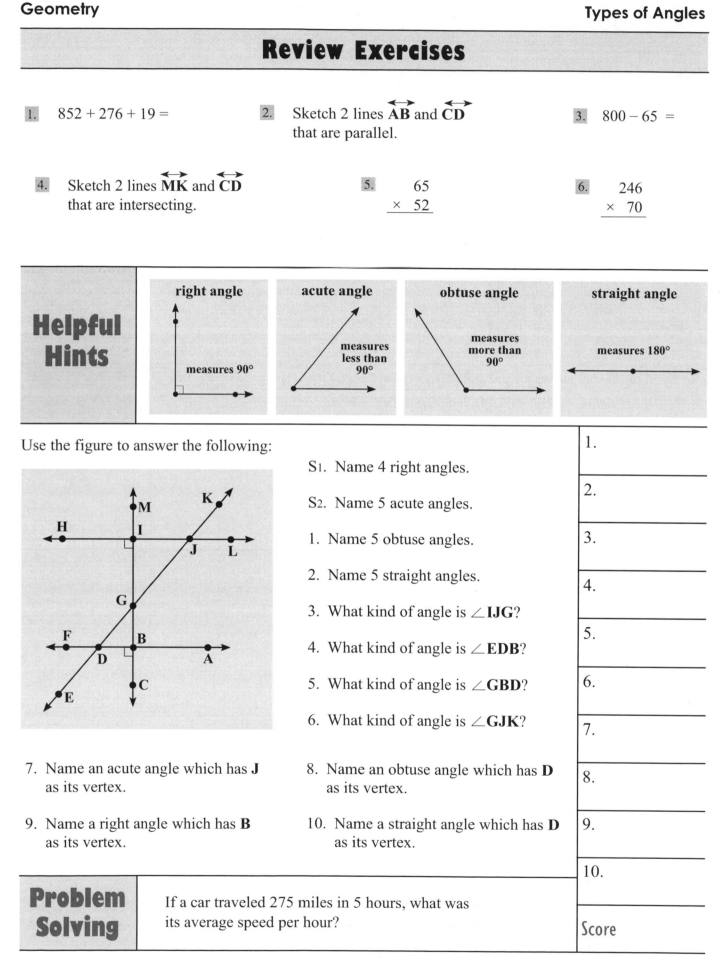

Helpful Hints

right angle	acute angle	obtuse angle	straight angle
measures 90°	measures less than 90°	measures more than 90°	measures 180°

Use the figure to answer the following:

S1. Name 4 right angles.

S2. Name 5 acute angles.

1. Name 5 obtuse angles.

2. Name 5 straight angles.

3. What kind of angle is $\angle$**IJG**?

4. What kind of angle is $\angle$**EDB**?

5. What kind of angle is $\angle$**GBD**?

6. What kind of angle is $\angle$**GJK**?

7. Name an acute angle which has **J** as its vertex.

8. Name an obtuse angle which has **D** as its vertex.

9. Name a right angle which has **B** as its vertex.

10. Name a straight angle which has **D** as its vertex.

1.
2.
3.
4.
5.
6.
7.
8.
9.
10.

Problem Solving

If a car traveled 275 miles in 5 hours, what was its average speed per hour?

Score

8

Review Exercises

1. $77 + 888 + 666 =$

2. $5,012 - 763 =$

3. $6 \times 108 =$

4. $\begin{array}{r} 600 \\ \times\ 32 \\ \hline \end{array}$

5. $\begin{array}{r} 365 \\ \times\ 402 \\ \hline \end{array}$

6. $8\overline{)8,008}$

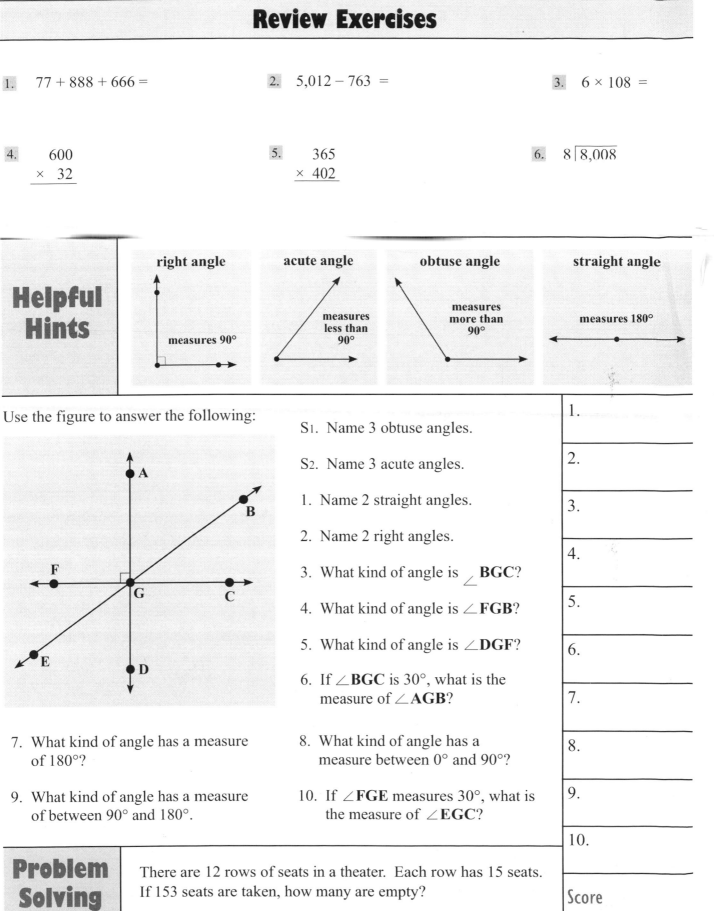

Helpful Hints

right angle
measures 90°

acute angle
measures less than 90°

obtuse angle
measures more than 90°

straight angle
measures 180°

Use the figure to answer the following:

S1. Name 3 obtuse angles.

S2. Name 3 acute angles.

1. Name 2 straight angles.

2. Name 2 right angles.

3. What kind of angle is $\angle$**BGC**?

4. What kind of angle is $\angle$**FGB**?

5. What kind of angle is $\angle$**DGF**?

6. If $\angle$**BGC** is 30°, what is the measure of $\angle$**AGB**?

7. What kind of angle has a measure of 180°?

8. What kind of angle has a measure between 0° and 90°?

9. What kind of angle has a measure of between 90° and 180°.

10. If $\angle$**FGE** measures 30°, what is the measure of $\angle$**EGC**?

1.

2.

3.

4.

5.

6.

7.

8.

9.

10.

Problem Solving

There are 12 rows of seats in a theater. Each row has 15 seats. If 153 seats are taken, how many are empty?

Score

Review Exercises

1. What kind of angle is ∠**BAC**?

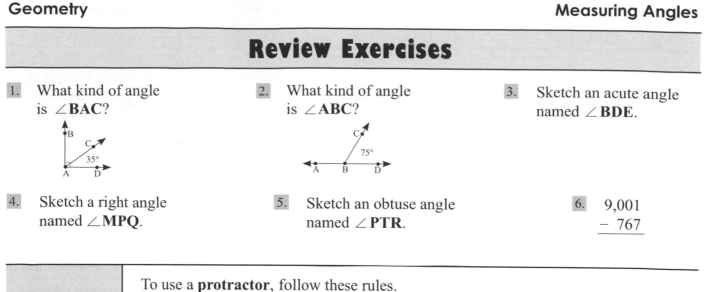

2. What kind of angle is ∠**ABC**?

3. Sketch an acute angle named ∠**BDE**.

4. Sketch a right angle named ∠**MPQ**.

5. Sketch an obtuse angle named ∠**PTR**.

6. 9,001
   − 767

To use a **protractor**, follow these rules.

1. Place the center point of the protractor on the vertex.
2. Place the zero mark on one edge of the angle.
3. Read the number where the other side of the angle crosses the protractor.
4. If the angle is acute, use the smaller number.
   If the angle is obtuse, use the larger number.

Use the figure to answer the questions. Classify the angle as right, acute, obtuse, or straight. Then tell how many degrees the angle measures.

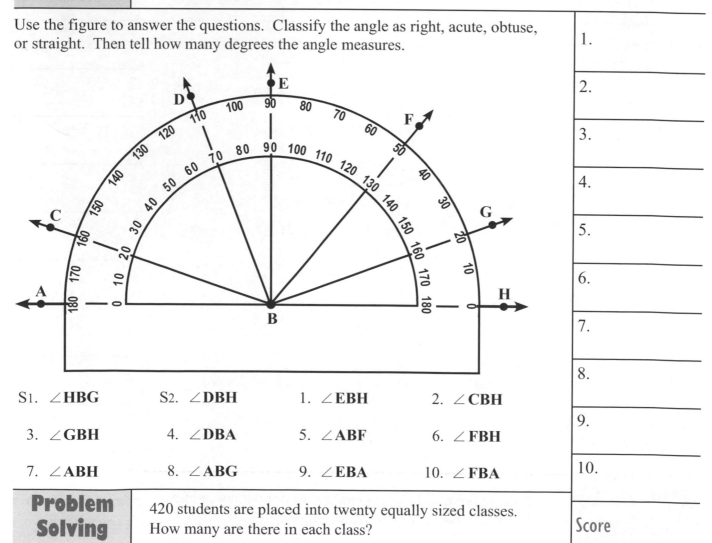

S1. ∠**HBG**	S2. ∠**DBH**	1. ∠**EBH**	2. ∠**CBH**
3. ∠**GBH**	4. ∠**DBA**	5. ∠**ABF**	6. ∠**FBH**
7. ∠**ABH**	8. ∠**ABG**	9. ∠**EBA**	10. ∠**FBA**

1.
2.
3.
4.
5.
6.
7.
8.
9.
10.

420 students are placed into twenty equally sized classes. How many are there in each class?

Score

10

Review Exercises

1. Sketch an obtuse angle named ∠**ABC**.

2. Sketch a right angle named ∠**EFG**.

3. Draw an acute angle named ∠**LMN**.

4. Draw two parallel lines $\overleftrightarrow{RS}$ and $\overleftrightarrow{GH}$.

5. What is the measure of angle ∠**BAC**?
   Hint:
   ∠**BAD** = 90°

6. What is the measure of angle ∠**TRW**?
   Hint:
   ∠**SRW** = 180°

Helpful Hints	Use what you have learned to answer the following questions.

With a protractor, measure the indicated angle in the figure. Tell the number of degrees. Also, classify the angle as acute, right, obtuse, or straight.

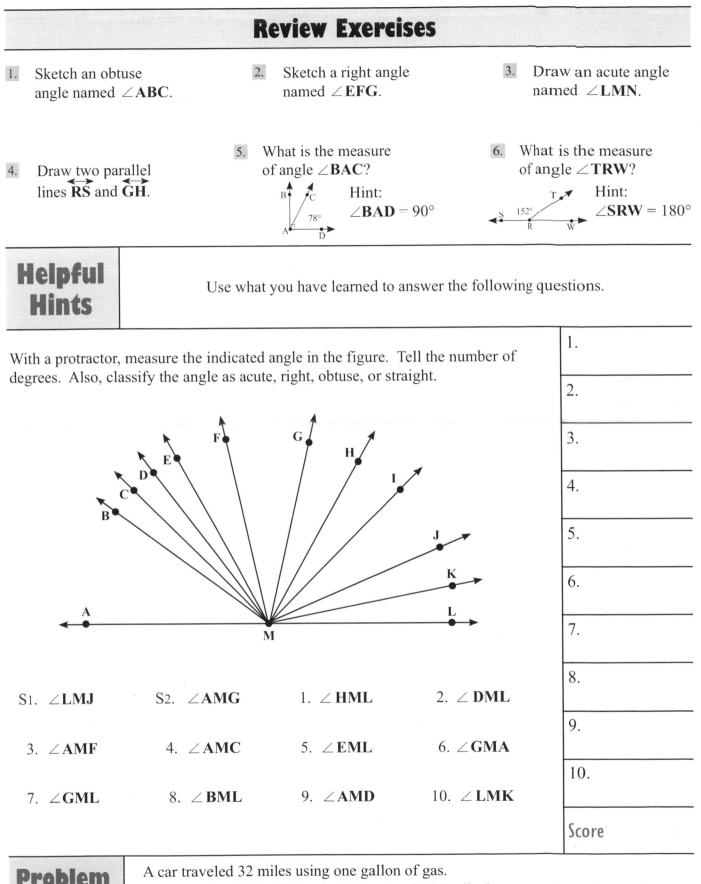

S1. ∠**LMJ**   S2. ∠**AMG**   1. ∠**HML**   2. ∠**DML**

3. ∠**AMF**   4. ∠**AMC**   5. ∠**EML**   6. ∠**GMA**

7. ∠**GML**   8. ∠**BML**   9. ∠**AMD**   10. ∠**LMK**

1.

2.

3.

4.

5.

6.

7.

8.

9.

10.

Score

Problem Solving	A car traveled 32 miles using one gallon of gas. How many gallons will be needed to travel 256 miles? If the cost is $3.00 per gallon, how much will it cost?

Review Exercises

1. What is the measure of angle ∠**ABC**?

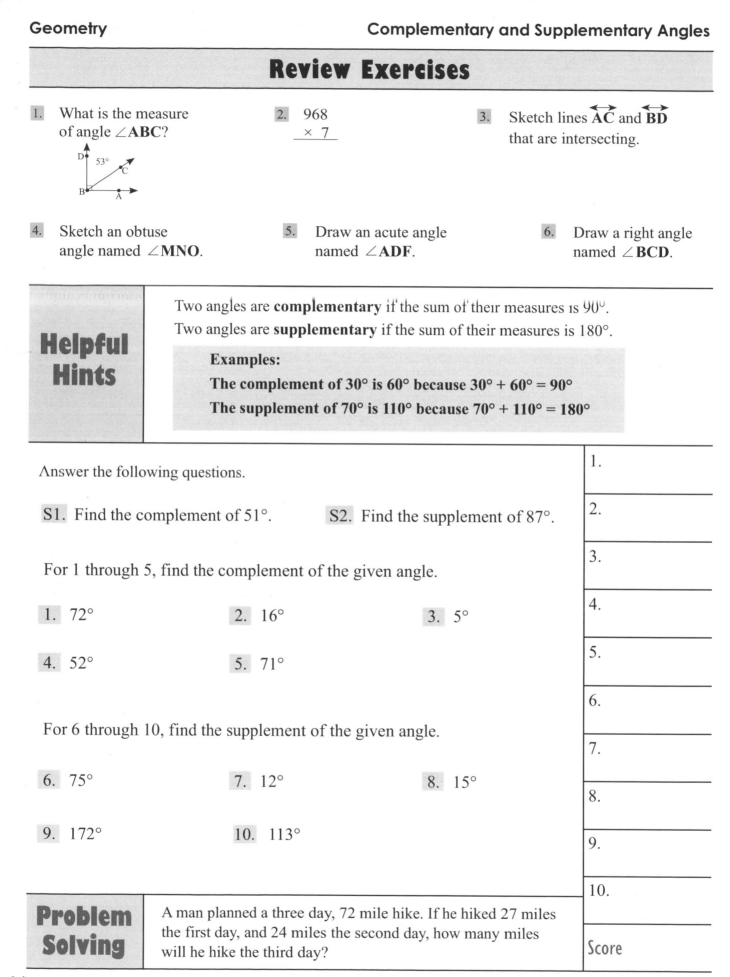

2. 968
   × 7

3. Sketch lines $\overleftrightarrow{AC}$ and $\overleftrightarrow{BD}$ that are intersecting.

4. Sketch an obtuse angle named ∠**MNO**.

5. Draw an acute angle named ∠**ADF**.

6. Draw a right angle named ∠**BCD**.

Helpful Hints

Two angles are **complementary** if the sum of their measures is 90°.

Two angles are **supplementary** if the sum of their measures is 180°.

Examples:

The complement of 30° is 60° because 30° + 60° = 90°

The supplement of 70° is 110° because 70° + 110° = 180°

Answer the following questions.

S1. Find the complement of 51°.   S2. Find the supplement of 87°.

For 1 through 5, find the complement of the given angle.

1. 72°   2. 16°   3. 5°

4. 52°   5. 71°

For 6 through 10, find the supplement of the given angle.

6. 75°   7. 12°   8. 15°

9. 172°   10. 113°

| 1. |
| 2. |
| 3. |
| 4. |
| 5. |
| 6. |
| 7. |
| 8. |
| 9. |
| 10. |
| Score |

Problem Solving

A man planned a three day, 72 mile hike. If he hiked 27 miles the first day, and 24 miles the second day, how many miles will he hike the third day?

14

Review Exercises

1. What is the supplement of 133° ?

2. What is the complement of 17° ?

3.
$$207 \\ \times \ \ 6$$

4. Draw two lines $\overleftrightarrow{AB}$ and $\overleftrightarrow{CD}$ that are parallel.

5.
$$736 \\ 75 \\ + \ \ 937$$

6.
$$7,112 \\ - \ 3,143$$

Helpful Hints

Use what you have learned to answer the following questions.

S1. Find the supplement of 65°.   S2. Find the compliment of 59°.

For 1 through 5, find the supplement of the given angle.

1. 79°   2. 141°   3. 33°

4. 19°   5. 62°

For 6 through 10, find the complement of the given angle.

6. 12°   7. 43°   8. 11°

9. 56°   10. 48°

1.	
2.	
3.	
4.	
5.	
6.	
7.	
8.	
9.	
10.	
Score	

Problem Solving

Each package of paper contains 500 sheets. How many sheets of paper are there in 16 packages?

Review Exercises

1. What is the complement of 14° ?

2. What is the supplement of 14° ?

3. What is the measure of angle ∠CDE?

    Hint: ∠CDE and ∠EDF are complementary angles.

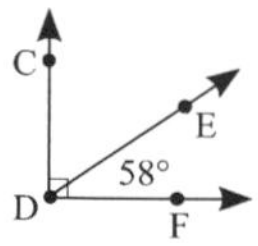

4. What is the measure of angle ∠LMN?

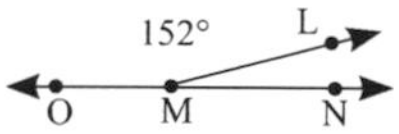

    Hint: ∠LMO and ∠LMN are supplementary angles.

5. 6⟌1596

6.   7,123
    −   368

Helpful Hints

To draw an angle follow these steps.
1. Draw a ray and put the center point of the protractor on the end point.
2. Align the ray with the base line of the protractor.
3. Locate the degree of the angle you wish to draw.
4. Place a dot at that point and connect it to the endpoint of the ray.

Example: Draw an angle with measure 60°.

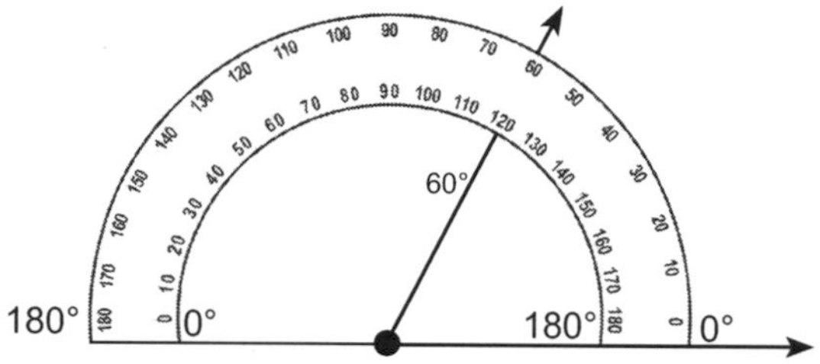

Use a protractor to draw angles with the following measures.

S1. 45°	S2. 120°	1. 30°
2. 110°	3. 70°	4. 52°
5. 90°	6. 160°	7. 40°

1.
2.
3.
4.
5.
6.
7.
8.
9.
10.

Problem Solving

If Lola earns $4,500 per month, what is her annual income? (Hint: How many months are in a year?)

Score

16

Review Exercises

1. What angle is supplementary to 82° ?

2. What angle is complementary to 17° ?

3.   367
      ×   15

4. What is the measure of angle ∠**DEF**?

5. What is the measure of angle ∠**HIJ?**

6. Draw two intersecting lines that intersect at point B.

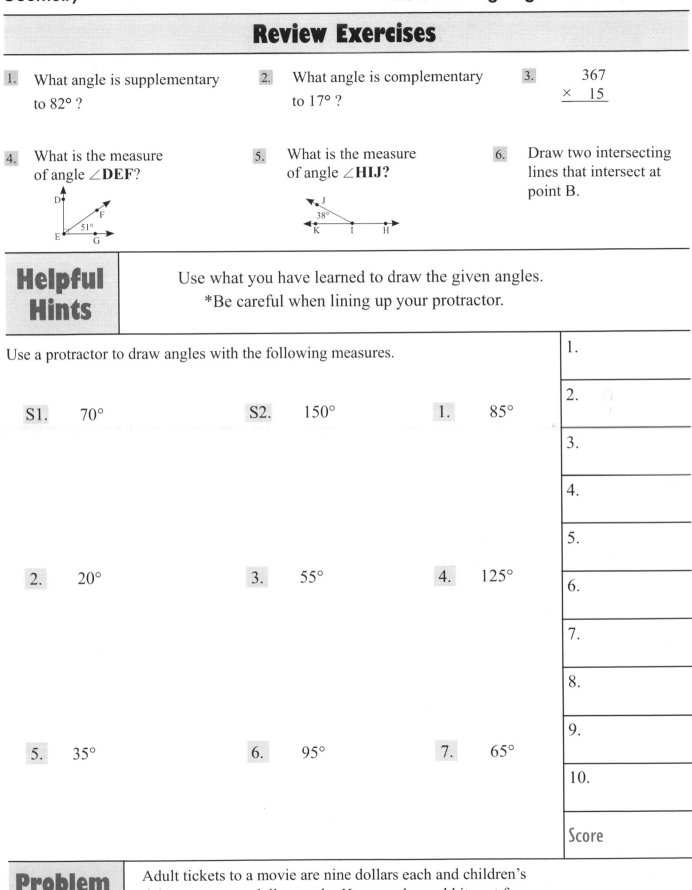

Helpful Hints

Use what you have learned to draw the given angles.
*Be careful when lining up your protractor.

Use a protractor to draw angles with the following measures.

S1. 70°	S2. 150°	1. 85°
2. 20°	3. 55°	4. 125°
5. 35°	6. 95°	7. 65°

1.

2.

3.

4.

5.

6.

7.

8.

9.

10.

Score

Problem Solving

Adult tickets to a movie are nine dollars each and children's tickets are seven dollars each. How much would it cost for two adult tickets and four children's tickets?

Review Exercises

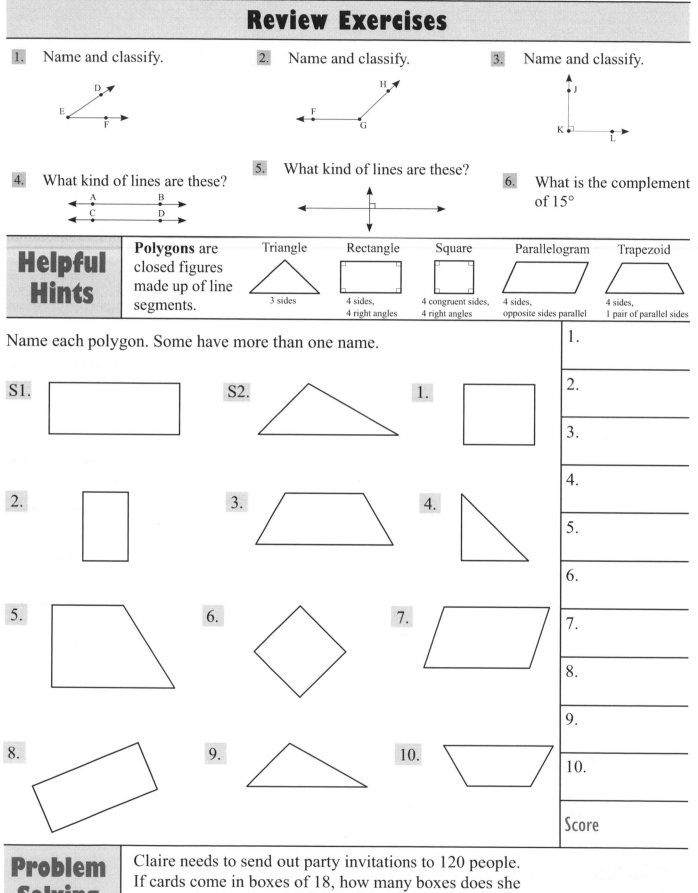

1. Name and classify.

2. Name and classify.

3. Name and classify.

4. What kind of lines are these?

5. What kind of lines are these?

6. What is the complement of 15°

Helpful Hints

Polygons are closed figures made up of line segments.

Triangle	Rectangle	Square	Parallelogram	Trapezoid
3 sides	4 sides, 4 right angles	4 congruent sides, 4 right angles	4 sides, opposite sides parallel	4 sides, 1 pair of parallel sides

Name each polygon. Some have more than one name.

S1.

S2.

1.

2.

3.

4.

5.

6.

7.

8.

9.

10.

1.

2.

3.

4.

5.

6.

7.

8.

9.

10.

Score

Problem Solving

Claire needs to send out party invitations to 120 people. If cards come in boxes of 18, how many boxes does she need to buy? How many cards will be left over?

Review Exercises

1. What is the complement of 14°?

2. What are the two names for this figure?

   5 in.

   5 in. [square] 5 in.

   5 in.

3. What is the supplement of 112°?

4. Sketch an obtuse angle named ∠ **BDF**.

5. Sketch a right angle named ∠ **LMN.**

6. Sketch 2 perpendicular lines meeting at point B.

Helpful Hints

Use what you have learned to answer the following questions.

* Some have more than one name.

Name each polygon. Some have more than one name.

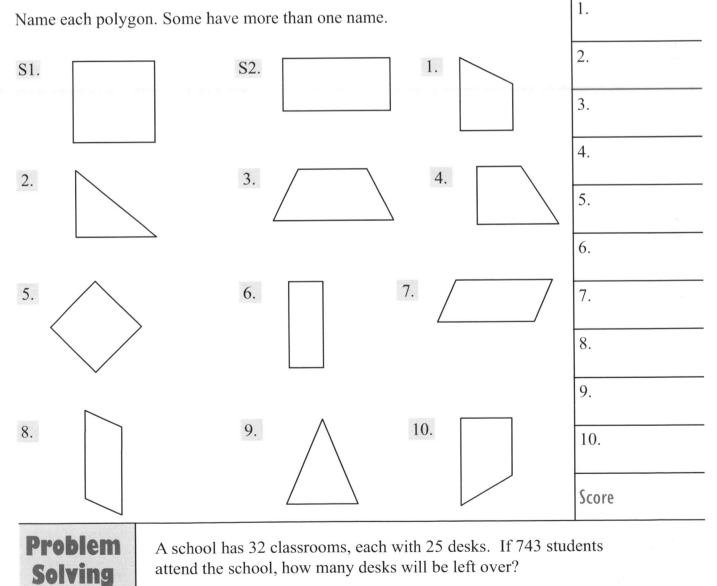

S1.

S2.

1.

2.

3.

4.

5.

6.

7.

8.

9.

10.

1.
2.
3.
4.
5.
6.
7.
8.
9.
10.
Score

Problem Solving

A school has 32 classrooms, each with 25 desks. If 743 students attend the school, how many desks will be left over?

Review Exercises

1. What polygon has three sides?

2. What polygon has one pair of parallel sides?

3. Name four polygons with four sides.

4. What polygons have 2 pairs of parallel sides?

5. What is the complement of 18°?

6. What is the supplement of 136°?

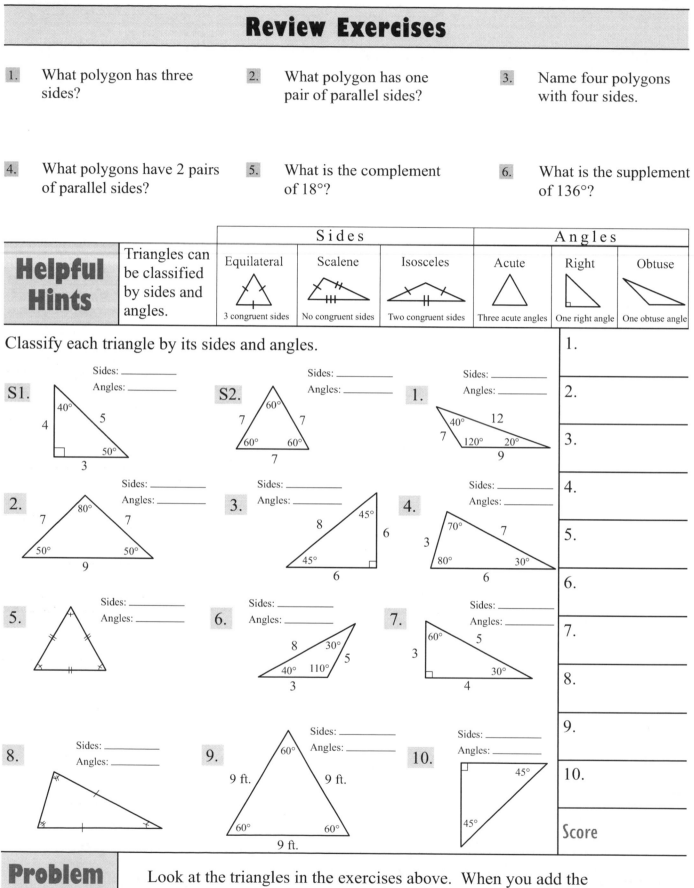

Helpful Hints

Triangles can be classified by sides and angles.

	Sides			Angles		
	Equilateral	Scalene	Isosceles	Acute	Right	Obtuse
	3 congruent sides	No congruent sides	Two congruent sides	Three acute angles	One right angle	One obtuse angle

Classify each triangle by its sides and angles.

S1. Sides: _____ Angles: _____

S2. Sides: _____ Angles: _____

1. Sides: _____ Angles: _____

2. Sides: _____ Angles: _____

3. Sides: _____ Angles: _____

4. Sides: _____ Angles: _____

5. Sides: _____ Angles: _____

6. Sides: _____ Angles: _____

7. Sides: _____ Angles: _____

8. Sides: _____ Angles: _____

9. Sides: _____ Angles: _____

10. Sides: _____ Angles: _____

1. _____
2. _____
3. _____
4. _____
5. _____
6. _____
7. _____
8. _____
9. _____
10. _____

Score _____

Problem Solving

Look at the triangles in the exercises above. When you add the three angles of any triangle, what will the total be?

Review Exercises

1. What kind of triangle has one obtuse angle?

2. What kind of triangle has three congruent sides?

3. What kind of triangle has all three sides of a different length?

4. What kind of triangle has two congruent sides?

5. If a triangle has three congruent sides what is the measure of each angle?

6. What kind of triangle has no congruent sides?

Use what you have learned to answer the following questions.

Classify each triangle by its side and angles using the facts given.
The facts are given in no particular order.

S1. Sides:_____
Angles:_____

Sides: 8, 8, 8
Angles: 60°, 60°, 60°

S2. Sides:_____
Angles:_____

Sides: 7, 8, 9
Angles: 70°, 60°, 50°

1. Sides:_____
Angles:_____

Sides: 7, 7, 10
Angles: 90°, 45°, 45°

2. Sides:_____
Angles:_____

Sides: 16, 10, 6
Angles: 30°, 110°, 40°

3. Sides:_____
Angles:_____

Sides: 12, 12, 12
Angles: 60°, 60°, 60°

4. Sides:_____
Angles:_____

Sides: 18, 14, 14
Angles: 80°, 50°, 50°

5. Sides:_____
Angles:_____

Sides: 3, 4, 5
Angles: 60°, 90°, 30°

6. Sides:_____
Angles:_____

Sides: 12, 16, 12
Angles: 45°, 45°, 90°

7. Sides:_____
Angles:_____

Sides: 9, 9, 9
Angles: 60°, 60°, 60°

8. Sides:_____
Angles:_____

Sides: 12, 6, 14
Angles: 30°, 70°, 80°

9. Sides:_____
Angles:_____

Sides: 6, 8, 8
Angles: 55°, 70°, 55°

10. Sides:_____
Angles:_____

Sides: 10, 8, 6
Angles: 40°, 90°, 50°

| 1. |
| 2. |
| 3. |
| 4. |
| 5. |
| 6. |
| 7. |
| 8. |
| 9. |
| 10. |
| Score |

If a triangle has two angles with measures of 74° and 65° what is the measure of the third angle?

21

Review Exercises

1. Classify by sides.

3 ft.   7 ft.
5 ft.

2. Classify by angles.

60°   30°

3. Classify by sides and angles.

80°
8 ft.   8 ft.
Sides: _____
Angles: _____   50°   50°
10 ft.

4.   756
   × 6

5.   6⟌936

6.   500 − 334 =

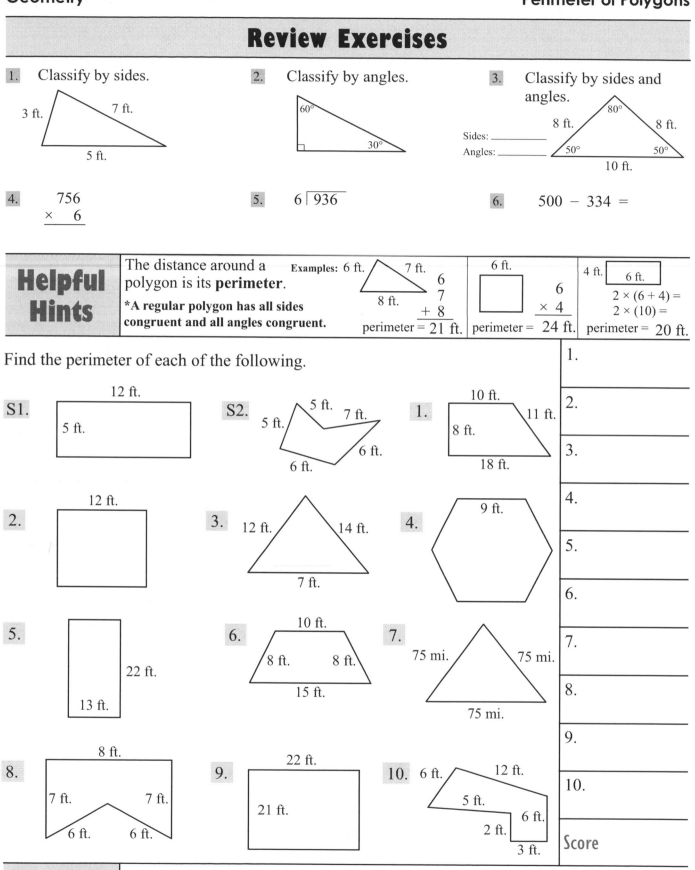

Helpful Hints

The distance around a polygon is its **perimeter**.

*A regular polygon has all sides congruent and all angles congruent.

Examples: 6 ft.   7 ft.
8 ft.
    6
    7
  + 8
perimeter = 21 ft.

6 ft.
    6
  × 4
perimeter = 24 ft.

4 ft.   6 ft.
2 × (6 + 4) =
2 × (10) =
perimeter = 20 ft.

Find the perimeter of each of the following.

S1.   12 ft.
5 ft.

S2.   5 ft.   7 ft.
5 ft.
6 ft.
6 ft.

1.   10 ft.   11 ft.
8 ft.
18 ft.

2.   12 ft.

3.   12 ft.   14 ft.
7 ft.

4.   9 ft.

5.   22 ft.
13 ft.

6.   10 ft.
8 ft.   8 ft.
15 ft.

7.   75 mi.   75 mi.
75 mi.

8.   8 ft.
7 ft.   7 ft.
6 ft.   6 ft.

9.   22 ft.
21 ft.

10.   6 ft.   12 ft.
5 ft.
2 ft.   6 ft.
3 ft.

1. _____
2. _____
3. _____
4. _____
5. _____
6. _____
7. _____
8. _____
9. _____
10. _____
Score

Problem Solving

What is the perimeter of a rectangle with a length of 15 ft. and a width of 12 ft. (Draw a sketch before solving the problem.)

Review Exercises

1. Classify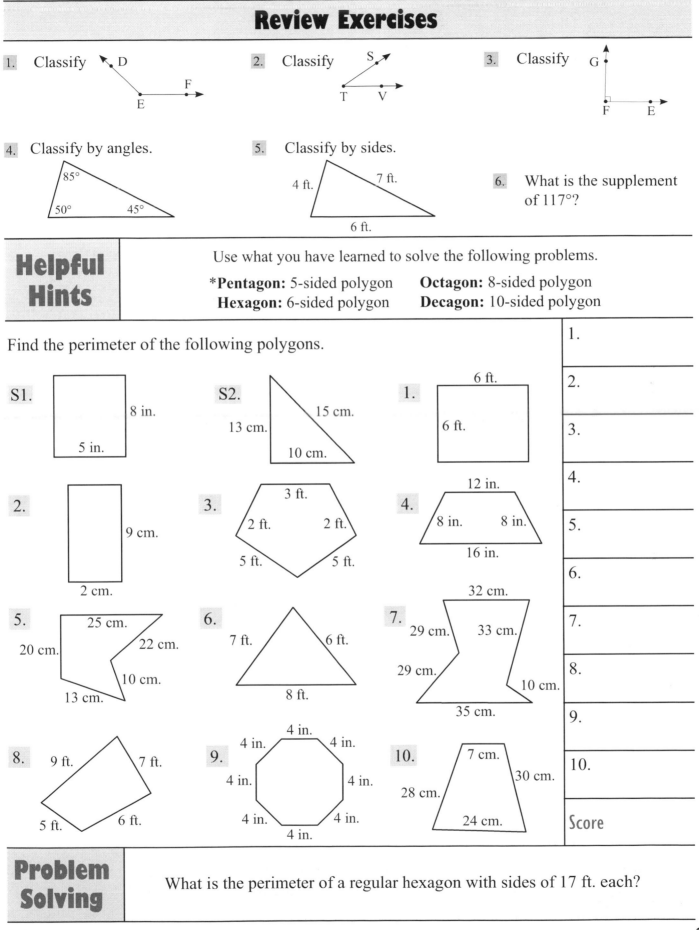

2. Classify

3. Classify

4. Classify by angles.

5. Classify by sides.

6. What is the supplement of 117°?

Helpful Hints

Use what you have learned to solve the following problems.

*Pentagon: 5-sided polygon   Octagon: 8-sided polygon
Hexagon: 6-sided polygon   Decagon: 10-sided polygon

Find the perimeter of the following polygons.

S1.   8 in.   5 in.

S2.   15 cm.   13 cm.   10 cm.

1.   6 ft.   6 ft.

2.   9 cm.   2 cm.

3.   3 ft.   2 ft.   2 ft.   5 ft.   5 ft.

4.   12 in.   8 in.   8 in.   16 in.

5.   25 cm.   20 cm.   22 cm.   10 cm.   13 cm.

6.   7 ft.   6 ft.   8 ft.

7.   32 cm.   29 cm.   33 cm.   29 cm.   10 cm.   35 cm.

8.   9 ft.   7 ft.   5 ft.   6 ft.

9.   4 in.   4 in.   4 in.   4 in.   4 in.   4 in.   4 in.   4 in.

10.   7 cm.   30 cm.   28 cm.   24 cm.

1.

2.

3.

4.

5.

6.

7.

8.

9.

10.

Score

Problem Solving

What is the perimeter of a regular hexagon with sides of 17 ft. each?

Review Exercises

1. Find the perimeter.

5 ft.

16 ft.

2. Find the perimeter.

19 in.

3. Find the perimeter of a triangle with sides 17 ft., 20 ft., and 23 ft.

4. Find the perimeter of a regular octagon with sides of 12 ft.

5. What is the third angle of a triangle with angles of 75° and 65°?

6. What angle is supplementary to 63°?

Helpful Hints

When solving problems related to perimeter follow these directions.
1. Read the problem carefully to fully understand what is asked.
2. Draw a sketch.
3. Solve the problem.

S1. Jim wants to build a fence around his yard. It is in the shape of a rectangle with a length of 32 ft. and a width of 18 ft. how many feet of fencing material does he need to buy?

S2. Find the perimeter of a regular decagon that has sides of 52 ft.

1. What is the perimeter of a square with sides of 96 ft.?

2. Jolie wants to put a wood frame around a painting that is in the shape of a rectangle. If the length is 36 inches and the width is 18 inches, how many inches of wood frame will be needed?

3. A square has a perimeter of 156 ft. What is the length of each side?

4. A banner is in the shape of an equilateral triangle. If each side is 57 inches, what is the perimeter of the banner?

5. The perimeter of a regular hexagon is 138 inches. What is the length of each side?

6. Bill's yard is in the shape of a square with sides of 15 ft. If he wants to build a fence around the yard and materials are 12 dollars per foot, how much will the fence cost?

7. An equilateral triangle has a perimeter of 291 inches. What is the length of each side?

1.

2.

3.

4.

5.

6.

7.

Score

Problem Solving

Three pounds of steak costs $14.97. What is the cost per pound?

Review Exercises

For 1-6 draw a sketch and find the perimeter.

1. **Square:**
    Sides 32 ft.

2. **Rectangle:**
    Length 35 ft.
    Width 17 ft.

3. **Regular pentagon:**
    Sides 17 ft.

4. **Equilateral triangle:**
    Sides 113 inches

5. **Regular Hexagon:**
    Sides 17 inches

6. **Rectangle:**
    Length 116 ft.
    Width 90 ft.

Helpful Hints

Use what you have learned to solve the following problems.
*Remember to draw a sketch.

S1. A rectangular garden is 17 ft. by 12 ft. How many feet of fencing is needed to build a fence around the garden?

S2. A kitchen floor is in the shape of a square with sides of 18 ft. Mrs. Smith wants to trim the sides of the floor with wood. If wood trim comes in sections three feet long, how many sections must she buy?

1. How many feet is it around a square garden with sides of 17 ft.?

2. If a regular octagon has a perimeter of 128 inches, how long is each side?

3. A ranch is in the shape of a square with sides 10 miles long. If a runner can travel 5 miles per hour, how long will it take to run around the ranch?

4. The perimeter of a rectangle is 60 ft. If the length is 18 ft., what is the width?

5. Find the perimeter of a regular decagon with sides of 19 ft.

6. The perimeter of a square is 504 inches. What is the length of each side?

7. The sum of two sides of an equilateral triangle is 96 ft. What are the lengths of the sides of the triangle?

1.
2.
3.
4.
5.
6.
7.

Score

Problem Solving

A plane traveled 3,600 miles in eight hours.
What was its average speed per hour?

Review Exercises

1. Classify by sides

7 ft. / 13 ft. / 12 ft.

2. Classify by angles

45° / 45°

3. What angle is complementary to 13° ?

4. Find the perimeter of a square with sides 27 ft.

5. Two angles in a triangle are 90° and 30°. What is the measure of the third angle?

6. The perimeter of a square is 64 ft. What is the length of each side?

Helpful Hints

These are the parts of a **circle**.

diameter   radius   chord   center

The length of the diameter is twice that of the radius.

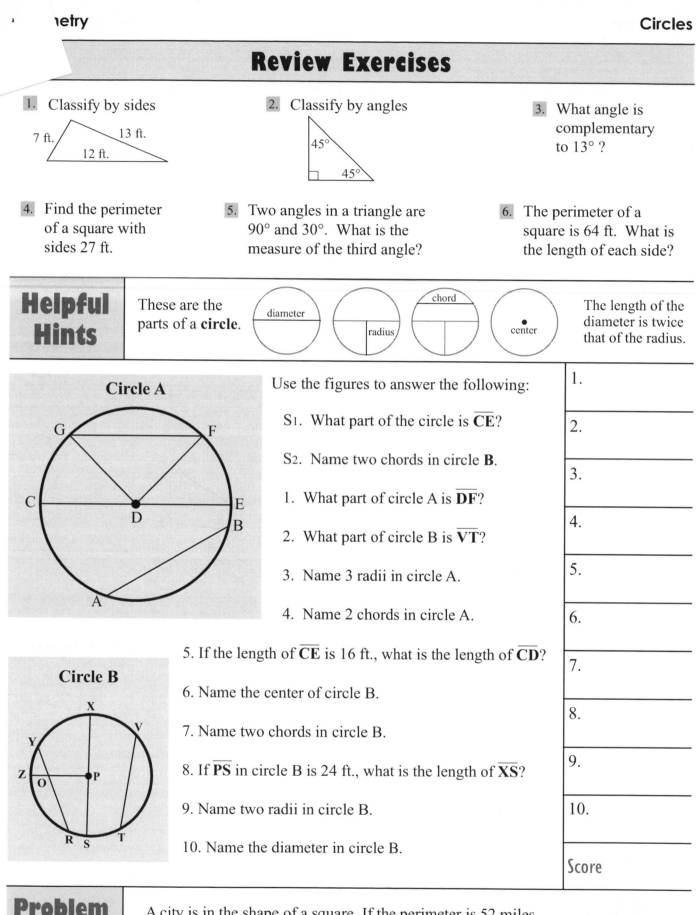

Circle A

Use the figures to answer the following:

S1. What part of the circle is $\overline{CE}$?

S2. Name two chords in circle **B**.

1. What part of circle A is $\overline{DF}$?

2. What part of circle B is $\overline{VT}$?

3. Name 3 radii in circle A.

4. Name 2 chords in circle A.

Circle B

5. If the length of $\overline{CE}$ is 16 ft., what is the length of $\overline{CD}$?

6. Name the center of circle B.

7. Name two chords in circle B.

8. If $\overline{PS}$ in circle B is 24 ft., what is the length of $\overline{XS}$?

9. Name two radii in circle B.

10. Name the diameter in circle B.

1.

2.

3.

4.

5.

6.

7.

8.

9.

10.

Score

Problem Solving

A city is in the shape of a square. If the perimeter is 52 miles, what is the length of each side of the city?

Review Exercises

1. If the radius of a circle is 16 ft., what is the length of the diameter?

2. The perimeter of a regular octagon is 72 ft. What is the length of each side?

3. 3.14
    × 6

4. $\frac{22}{7} \times 21 =$

5. Find the missing angle.

6. Classify sides _____
   angles _____

| **Helpful Hints** | Use what you have learned to solve the following questions. |

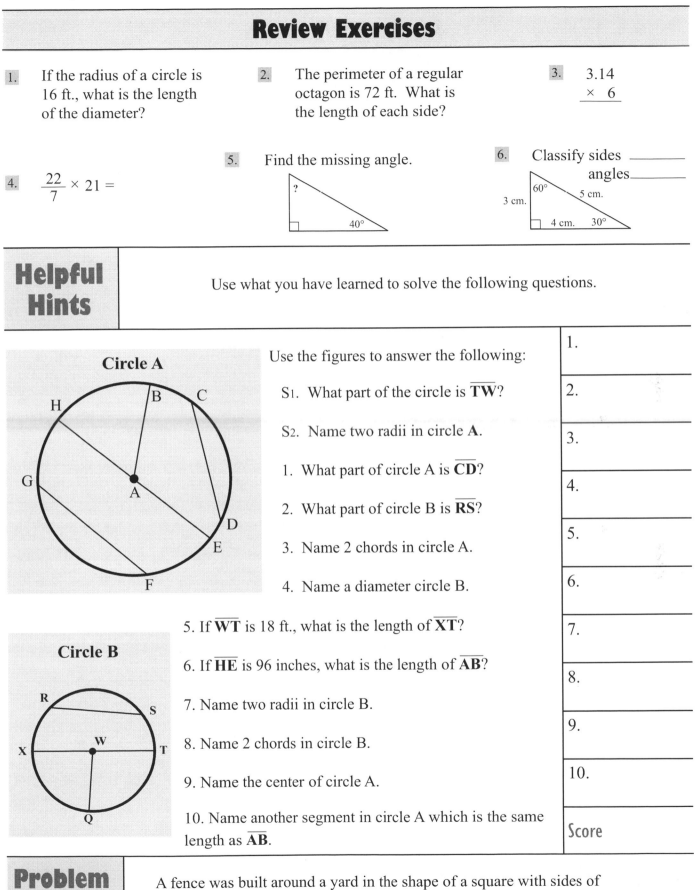

Circle A

Use the figures to answer the following:

S1. What part of the circle is $\overline{TW}$?

S2. Name two radii in circle **A**.

1. What part of circle A is $\overline{CD}$?

2. What part of circle B is $\overline{RS}$?

3. Name 2 chords in circle A.

4. Name a diameter circle B.

5. If $\overline{WT}$ is 18 ft., what is the length of $\overline{XT}$?

6. If $\overline{HE}$ is 96 inches, what is the length of $\overline{AB}$?

Circle B

7. Name two radii in circle B.

8. Name 2 chords in circle B.

9. Name the center of circle A.

10. Name another segment in circle A which is the same length as $\overline{AB}$.

1.

2.

3.

4.

5.

6.

7.

8.

9.

10.

Score

| **Problem Solving** | A fence was built around a yard in the shape of a square with sides of 14 ft. If materials cost $12 per foot, how much did the fence cost? |

Review Exercises

1. If the radius of a circle is 12 ft., what is the length of the diameter?

2. $\dfrac{22}{7} \times 35 =$

3. $\begin{array}{r} 3.14 \\ \times\ \ \ 8 \\ \hline \end{array}$

4. The diameter of a circle is 56 inches. What is the length of the radius?

5. What kind of triangle has all sides of a different length?

6. What kind of triangle has two sides the same length?

Helpful Hints

The distance around a circle is called its **circumference**. The Greek letter π = pi = 3.14 or $\dfrac{22}{7}$. To find the **circumference**, multiply $\pi \times$ **diameter**. **Circumference = $\pi \times$ d**

Examples:

$C = \pi \times d$
$C = 3.14 \times 6$

6 ft.

$\begin{array}{r} 3.14 \\ \times\ \ \ 6 \\ \hline 18.84\ \text{ft.} \end{array}$

7 ft.

$C = \pi \times d$
$C = \dfrac{22}{\cancel{7}^{1}} \times \dfrac{\cancel{14}^{2}}{1} = 44 \text{ ft.}$

(Hint: If the diameter is divisible by 7, use $\pi = \dfrac{22}{7}$)

Find the circumference of each of the following.

S1. 4 ft.

S2. 7 ft.

1. 6 ft.

2. 4 ft.

3. A circle with diameter 9 ft.

4. A circle with radius 14 ft.

5. 12 ft.

6. 5 ft.

7. A circle with radius 2 ft.

1. _____

2. _____

3. _____

4. _____

5. _____

6. _____

7. _____

Score

Problem Solving

Mr. Vargas earned $67,200 last year. What were his average monthly earnings?

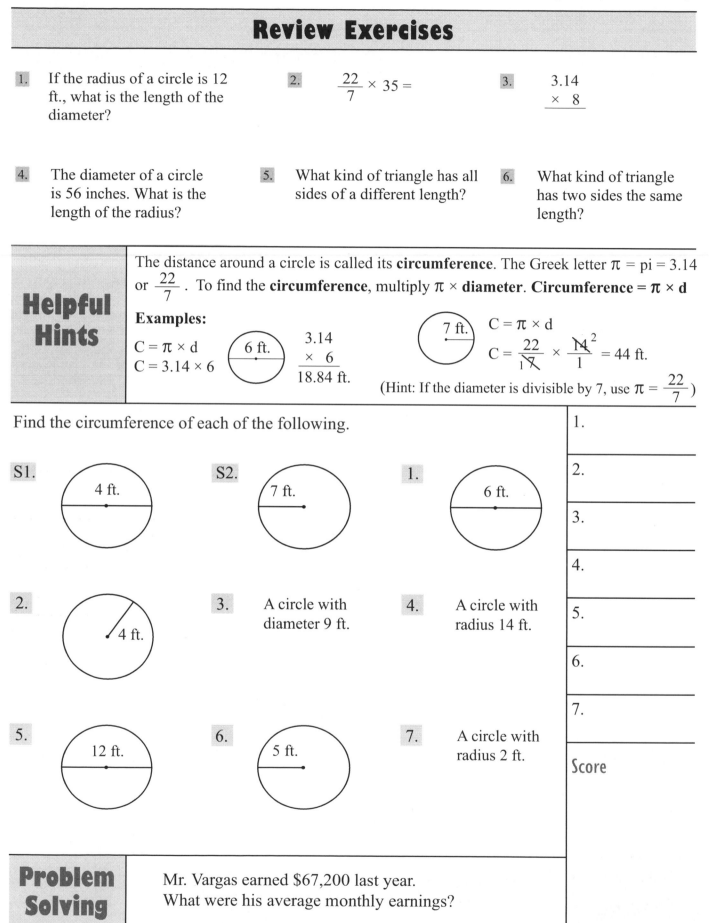

Review Exercises

For 1-6, find the perimeter for each figure. Draw a sketch.

1. A square with sides of 73 inches.

2. A rectangle with length 32 ft., and width 18 ft.

3. A regular pentagon with sides 63 meters.

4. An equilateral triangle with sides 29 ft.

5. A regular octagon with sides 46 ft.

6. A regular hexagon with sides 112 ft.

Helpful Hints

Use what you have learned to solve the following problems.
1. If there is no figure, draw a sketch.
2. If the diameter is divisible by 7, use $\pi = \dfrac{22}{7}$

Find the circumference of each of the following.

S1. 14 ft.

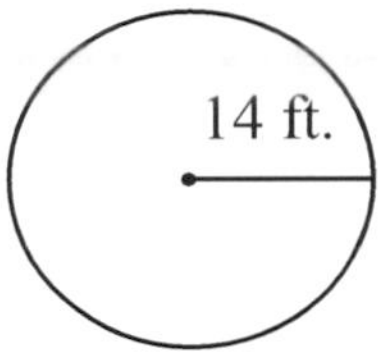

S2. 9 ft.

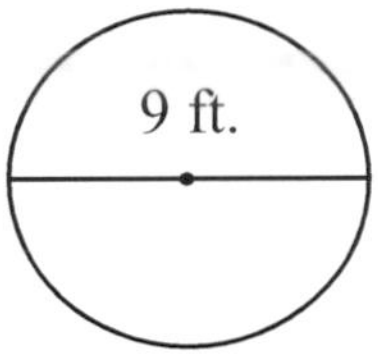

1. 15 ft.

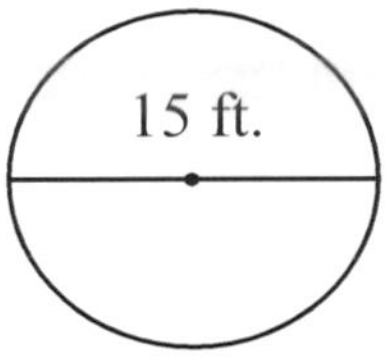

2. 8 ft.

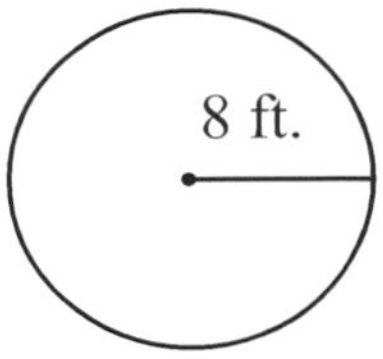

3. 21 ft.

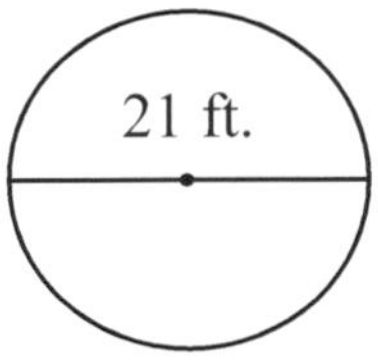

4. A circle with radius 21 ft.

5. A circle with diameter 35 ft.

6. 6 ft.

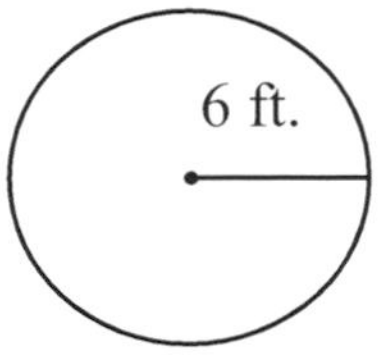

7. 100 ft.

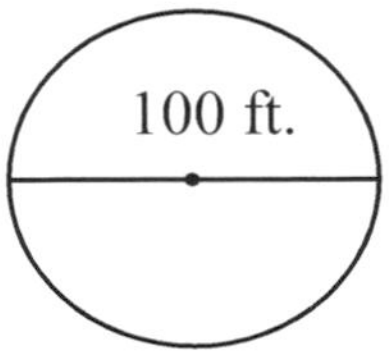

1. _____

2. _____

3. _____

4. _____

5. _____

6. _____

7. _____

Score

Problem Solving

A fountain is in the shape of a circle. If its radius is 14 ft., how far is it all the way around the fountain?

Review Exercises

1. Find the circumference.

14 ft.

2. Find the perimeter.

12 ft.
33 ft.

3. Find the circumference

14 ft.

4. $3.14 \overline{)25.12}$

5. $66 \div \dfrac{22}{7} =$

6. Classify by angles.

35°
123°   22°

| **Helpful Hints** | Use what you have learned to solve the following problems.
1. Read the problem carefully to understand what is being asked, draw a sketch.
2. If the diameter is divisible by 7, use $\pi = \dfrac{22}{7}$ * diameter = C ÷ π |

S1. A circular dodge ball court has a diameter of 35 ft. How far is it all the way around the court?

S2. Mrs. Isaac is making cloth emblems for her club's jackets. If each emblem has a radius of 7 cm., what is the circumference of each emblem?

1. Tara has a frisbee with a diameter of 14 inches. What is the circumference of the frisbee?

2. A city is surrounded by a road in the shape of a circle. If the radius of the road is 14 miles, how far would you drive to make a complete circle around the city?

3. Circle A has diameter 21 ft. and Circle B has diameter 28 ft. How much greater is the circumference of Circle B than Circle A?

4. The circumference of a circle is 12.56 ft. What is the diameter? (Use π = 3.14)

5. The circumference of a circle is 18.84 inches. What is the radius? (Use π = 3.14)

1.

2.

3.

4.

5.

6.

7.

Score

6. Find the circumference.

$3\frac{1}{2}$ ft.

7. Find the circumference.

10 ft.

Problem Solving

The perimeter of a square is 496 ft. What is the length of each side?

Review Exercises

1. What is the complement of 23°?

2. What is the supplement of 97°?

3. What polygon has one pair of parallel sides?

4. $88 \div \dfrac{22}{7} =$

5. $3.14\overline{)37.68}$

6. Two angles of a triangle are 37° and 72°. What is the measure of the third angle?

Helpful Hints

Use what you have learned to solve the following problems.

* Draw a sketch.
* Decide which is easier, using $\pi = 3.14$ or $\pi = \dfrac{22}{7}$ * $d = C \div \pi$

S1. The circumference of a circle is 110 ft. What is the diameter? (Use $\pi = \dfrac{22}{7}$)

S2. The circumference of a circle is 47.1 ft. What is the diameter?

1. The radius of a circular shaped fountain is 14 ft. What is the diameter of the fountain?

2. A race track in the shape of a circle has a radius of 56 meters. What is the circumference of the race track?

3. From the north pole through the center of the earth to the south pole is approximately 8,000 miles. What is the distance around the equator?

4. If the circumference of a circle is 264 ft., what is the diameter? ($\pi = \dfrac{22}{7}$)

5. If the circumference of a circle is 62.8 ft., what is the diameter? ($\pi = 3.14$)

6. Find the circumference.

5 ft.

7. Find the circumference.

56 ft.

1.
2.
3.
4.
5.
6.
7.
Score

Problem Solving

A sports court is in the shape of a rectangle with length 84 ft. and width 36 ft. What is the perimeter of the sports court?

Review Exercises

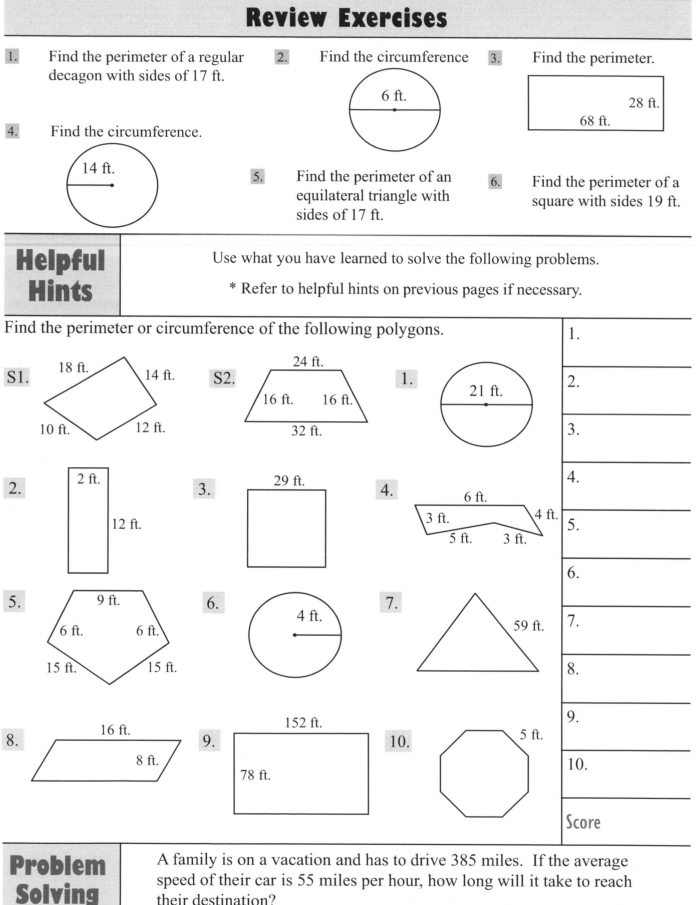

1. Find the perimeter of a regular decagon with sides of 17 ft.

2. Find the circumference

6 ft.

3. Find the perimeter.

28 ft.
68 ft.

4. Find the circumference.

14 ft.

5. Find the perimeter of an equilateral triangle with sides of 17 ft.

6. Find the perimeter of a square with sides 19 ft.

Helpful Hints

Use what you have learned to solve the following problems.

* Refer to helpful hints on previous pages if necessary.

Find the perimeter or circumference of the following polygons.

S1. 18 ft. 14 ft. 10 ft. 12 ft.

S2. 24 ft. 16 ft. 16 ft. 32 ft.

1. 21 ft.

2. 2 ft. 12 ft.

3. 29 ft.

4. 6 ft. 3 ft. 4 ft. 5 ft. 3 ft.

5. 9 ft. 6 ft. 6 ft. 15 ft. 15 ft.

6. 4 ft.

7. 59 ft.

8. 16 ft. 8 ft.

9. 152 ft. 78 ft.

10. 5 ft.

1.
2.
3.
4.
5.
6.
7.
8.
9.
10.
Score

Problem Solving

A family is on a vacation and has to drive 385 miles. If the average speed of their car is 55 miles per hour, how long will it take to reach their destination?

Review Exercises

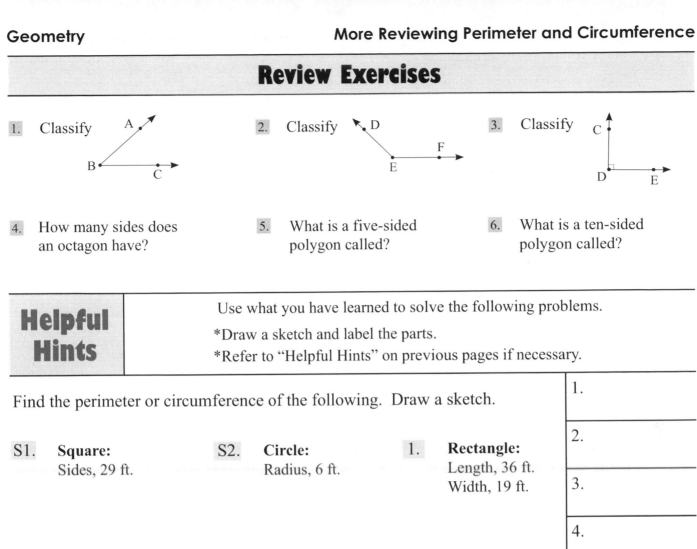

1. Classify

2. Classify

3. Classify

4. How many sides does an octagon have?

5. What is a five-sided polygon called?

6. What is a ten-sided polygon called?

Helpful Hints

Use what you have learned to solve the following problems.

*Draw a sketch and label the parts.

*Refer to "Helpful Hints" on previous pages if necessary.

Find the perimeter or circumference of the following. Draw a sketch.

S1. **Square:**
Sides, 29 ft.

S2. **Circle:**
Radius, 6 ft.

1. **Rectangle:**
Length, 36 ft.
Width, 19 ft.

2. **Equilateral Triangle:**
Sides, 86 ft.

3. **Circle:**
Diameter, 28 ft.

4. **Regular Pentagon:**
Sides, 37 ft.

5. **Square:**
Sides, 115 ft.

6. **Regular Decagon:**
Sides, 39 ft.

7. **Rectangle:**
Length, 46 ft.
Width, 96 ft.

8. **Circle:**
Radius, 21 ft.

9. **Regular Hexagon:**
Sides, 42 ft.

10. If the perimeter of a square is 528 ft., what are the lengths of the sides?

1.

2.

3.

4.

5.

6.

7.

8.

9.

10.

Score

Problem Solving

During a week Julio started with 550 dollars. First he spent 136 dollars, then he earned 208 dollars, and finally he spent 316 dollars. How much money did he have left?

Review Exercises

1. Draw two lines $\overleftrightarrow{CD}$ and $\overleftrightarrow{FG}$, that are perpendicular.

2. Draw two lines $\overleftrightarrow{LM}$ and $\overleftrightarrow{RS}$, that are parallel.

3. What angle is complementary to 15°?

4. What angle is supplementary to 15°?

5. In a right isosceles triangle, one angle measures 90°. What are the measures of the other two angles?

6. What is the measure of ∠ **ABC**?

Helpful Hints

Use what you have learned to solve the following problems.
*Draw a sketch.
*Refer to previous pages if necessary.

S1. What is the circumference of a circular sports court if the diameter is 35 ft?

S2. A flower garden is in the shape of a regular pentagon. If its sides are 17 ft., what is the perimeter of the flower garden?

1. A coach wants to paint a white line around the perimeter of a rectangular court. If the width is 35 ft. and the length is 70 ft., what is the perimeter of the court?

2. The circumference of a circle is 154 ft. What is the diameter? $(\pi = \frac{22}{7})$

3. The sides of a square are 180 ft. What is the perimeter?

4. The perimeter of a regular hexagon is 228 ft. What are the lengths of the sides?

5. A man wants to build a fence around a yard in the shape of a square with sides of 24 ft. If fencing costs 15 dollars per foot, what will be the cost of the fence?

6. The perimeter of a square is 624 ft. What are the lengths of the sides?

7. What is the perimeter of a regular octagon with sides of 23 ft.?

1.	
2.	
3.	
4.	
5.	
6.	
7.	
Score	

Problem Solving

A tank holds 55,000 gallons. If 12,500 gallons were removed one day, and 13,450 gallons the next day, how many gallons are left?

Review Exercises

For 1-6, find the perimeter or circumference. (Draw a sketch.)

1. **Square:**
    Sides, 16 ft.

2. **Rectangle:**
    Length, 26 ft.
    Width, 19 ft.

3. **Circle:**
    Radius, 3 ft.

4. **Circle:**
    Diameter, 21 ft.

5. **Regular Pentagon:**
    Sides, 28 ft.

6. **Equilateral Triangle:**
    Sides, 33 ft.

Helpful Hints

Use what you have learned to solve the following problems.

S1. What is the diameter of a circle with circumference 198 ft.? ($\pi = \frac{22}{7}$)

S2. What is the radius of a circle with circumference 18.84 ft.? ($\pi = 3.14$)

1. A circular race track has a circumference of 3 miles. In a 375 mile car race, how many laps around the track will each auto travel?

2. The walls of the Pentagon Building are 921 ft. long. What is the perimeter of the building? (Hint: The building is in the shape of a regular pentagon.)

3. Each side of a square is 4 ft. long. How many inches is it all the way around the square?

4. The perimeter of a rectangle is 60 ft. If the length is 18 ft., what is the width?

5. The diameter of a track is 7 miles. How many laps must a car drive to cover a distance of 88 miles?

6. What is the perimeter of a regular octagon with sides 38 ft.?

7. Circle A has a diameter of 14 ft. Circle B has a diameter of 35 ft. How much longer is the circumference of Circle B than Circle A?

1.

2.

3.

4.

5.

6.

7.

Score

Problem Solving

A family planned a 750 mile trip. They drove 335 miles the first day and 255 miles the second day. How many more miles must they drive?

35

Review Exercises

1. List three types of polygons.

2. List the four types of angles.

3. List four parts of a circle.

4. List the three types of triangles classified by sides.

5. List three types of triangles classified by angles.

6. List four polygons that have at least five sides.

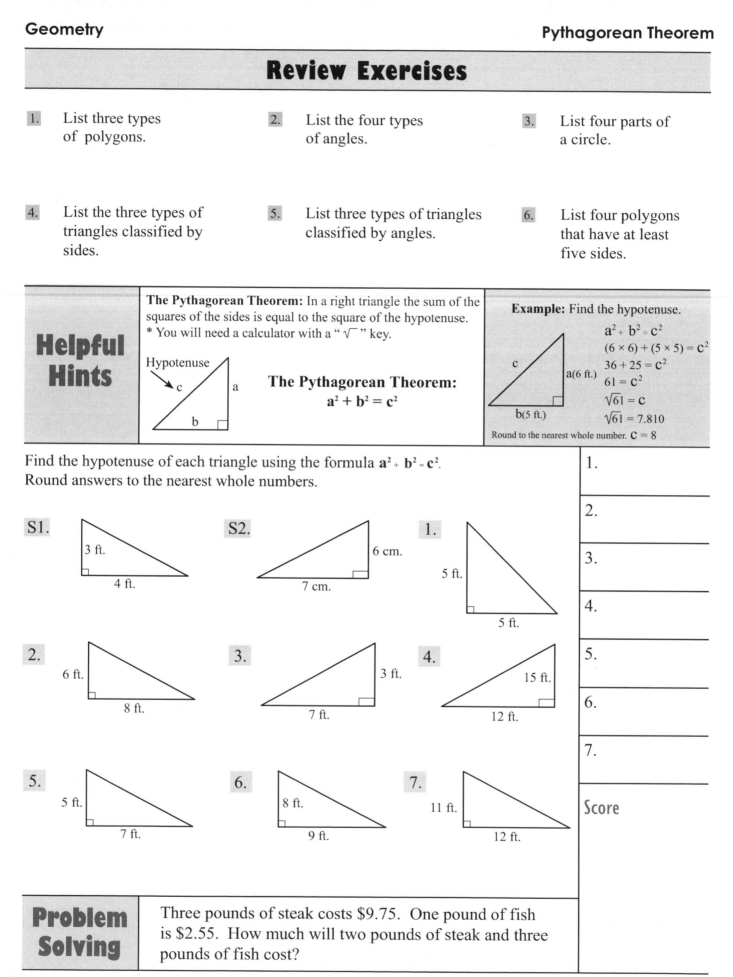

Helpful Hints

The Pythagorean Theorem: In a right triangle the sum of the squares of the sides is equal to the square of the hypotenuse.
* You will need a calculator with a " $\sqrt{}$ " key.

Hypotenuse

The Pythagorean Theorem:
$$a^2 + b^2 = c^2$$

Example: Find the hypotenuse.

$a^2 + b^2 = c^2$
$(6 \times 6) + (5 \times 5) = c^2$
$36 + 25 = c^2$
$61 = c^2$
$\sqrt{61} = c$
$\sqrt{61} = 7.810$

Round to the nearest whole number. **c = 8**

Find the hypotenuse of each triangle using the formula $a^2 + b^2 = c^2$.
Round answers to the nearest whole numbers.

S1.   3 ft.   4 ft.

S2.   6 cm.   7 cm.

1.   5 ft.   5 ft.

2.   6 ft.   8 ft.

3.   3 ft.   7 ft.

4.   15 ft.   12 ft.

5.   5 ft.   7 ft.

6.   8 ft.   9 ft.

7.   11 ft.   12 ft.

1. ____

2. ____

3. ____

4. ____

5. ____

6. ____

7. ____

Score

Problem Solving

Three pounds of steak costs $9.75. One pound of fish is $2.55. How much will two pounds of steak and three pounds of fish cost?

36

Review Exercises

For 1-6, find the perimeter or circumference. (Draw a sketch.)

1. **Square:**
    Sides, 17 ft.

2. **Rectangle:**
    Length, 19 ft.
    Width, 7 ft.

3. **Regular Hexagon:**
    Sides, 29 ft.

4. **Circle:**
    Radius, 6 ft.

5. **Circle:**
    Diameter, 7 ft.

6. **Equilateral Triangle:**
    Sides, 175 ft.

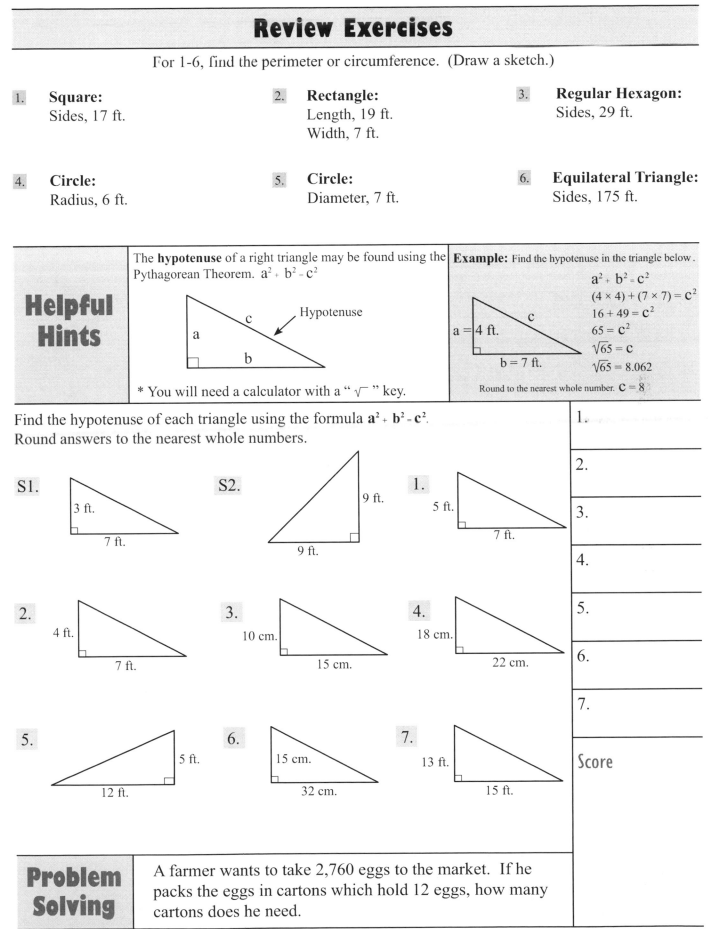

Helpful Hints

The **hypotenuse** of a right triangle may be found using the Pythagorean Theorem. $a^2 + b^2 = c^2$

Hypotenuse

* You will need a calculator with a " $\sqrt{\ }$ " key.

Example: Find the hypotenuse in the triangle below.

$$a^2 + b^2 = c^2$$
$$(4 \times 4) + (7 \times 7) = c^2$$
$$16 + 49 = c^2$$
$$65 = c^2$$
$$\sqrt{65} = c$$
$$\sqrt{65} = 8.062$$

$a = 4$ ft. $b = 7$ ft.

Round to the nearest whole number. $c = 8$

Find the hypotenuse of each triangle using the formula $a^2 + b^2 = c^2$.
Round answers to the nearest whole numbers.

S1.   3 ft.   7 ft.

S2.   9 ft.   9 ft.

1.   5 ft.   7 ft.

2.   4 ft.   7 ft.

3.   10 cm.   15 cm.

4.   18 cm.   22 cm.

5.   5 ft.   12 ft.

6.   15 cm.   32 cm.

7.   13 ft.   15 ft.

1.
_____

2.
_____

3.
_____

4.
_____

5.
_____

6.
_____

7.
_____

Score

Problem Solving

A farmer wants to take 2,760 eggs to the market. If he packs the eggs in cartons which hold 12 eggs, how many cartons does he need.

Review Exercises

1. Find the perimeter.

9 ft. [ ]
25 ft.

2. Find the perimeter.

[ ] 119 ft.

3. Find the perimeter.

(42 ft.)

4. Find the circumference.

(4 ft.)

5. How many sides does a hexagon have?

6. How many sides does an octagon have?

Helpful Hints

Area can be thought of as the amount of surface covered by an object. It is expressed in square units. To estimate area:
1. Count the number of squares completely shaded.
2. Count the number of squares more than half shaded.
3. Add the two numbers.

Example:

$$6$$
$$+\ 3$$
$$\overline{9}$$

The area is about 9 ft.^{2}

Each square is 1 ft.^{2}
ft.^{2} = square feet

Estimate the area of each shaded figure. Each square is 1 ft.^{2}.

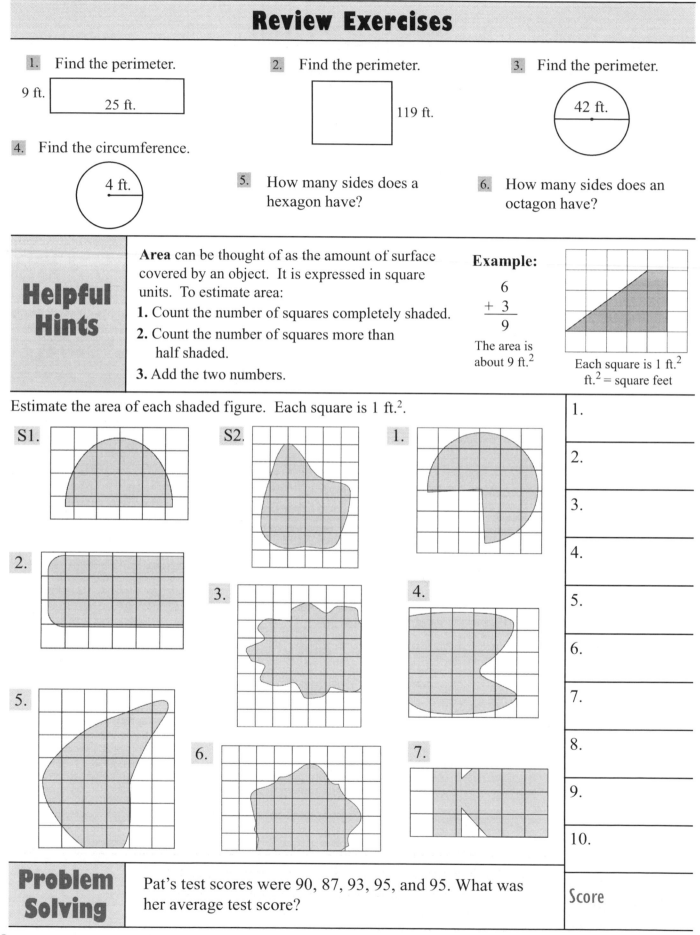

S1.

S2.

1.

2.

3.

4.

5.

6.

7.

1.
2.
3.
4.
5.
6.
7.
8.
9.
10.

Problem Solving

Pat's test scores were 90, 87, 93, 95, and 95. What was her average test score?

Score

Review Exercises

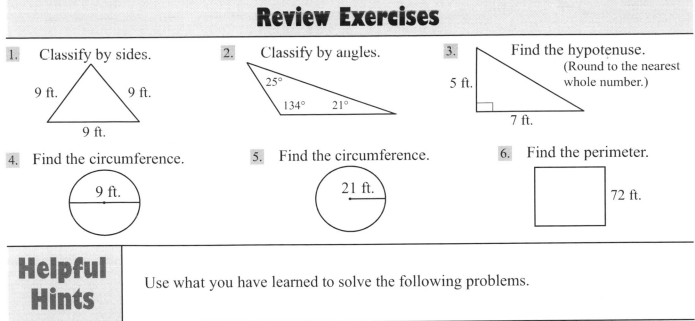

1. Classify by sides.

9 ft.   9 ft.

9 ft.

2. Classify by angles.

25°

134°   21°

3. Find the hypotenuse.
(Round to the nearest whole number.)

5 ft.

7 ft.

4. Find the circumference.

9 ft.

5. Find the circumference.

21 ft.

6. Find the perimeter.

72 ft.

Helpful Hints

Use what you have learned to solve the following problems.

Shade in the figure the area given. Each square is 1 ft.²

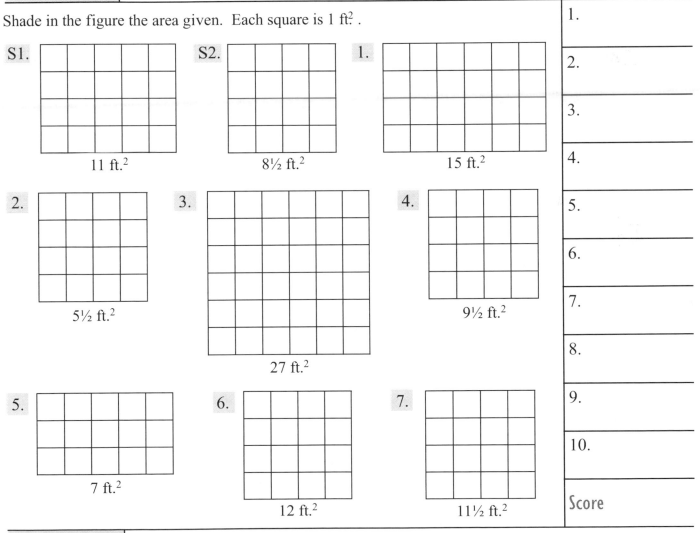

S1.

11 ft.²

S2.

8½ ft.²

1.

15 ft.²

2.

5½ ft.²

3.

27 ft.²

4.

9½ ft.²

5.

7 ft.²

6.

12 ft.²

7.

11½ ft.²

1.
2.
3.
4.
5.
6.
7.
8.
9.
10.

Score

Problem Solving

Sam bought a new car. He made a down payment of 2,000 dollars, and will make 48 monthly payments of 400 dollars. What is the total cost of the car?

Review Exercises

1. What is the complement of 6°?

2. What is the supplement of 17°?

3. Name three polygons with two pairs of parallel sides.

4. If the perimeter of a square is 580 ft., find the length of its sides.

5. Find the perimeter.

    16 ft.

6. Find the circumference.

    2 ft.

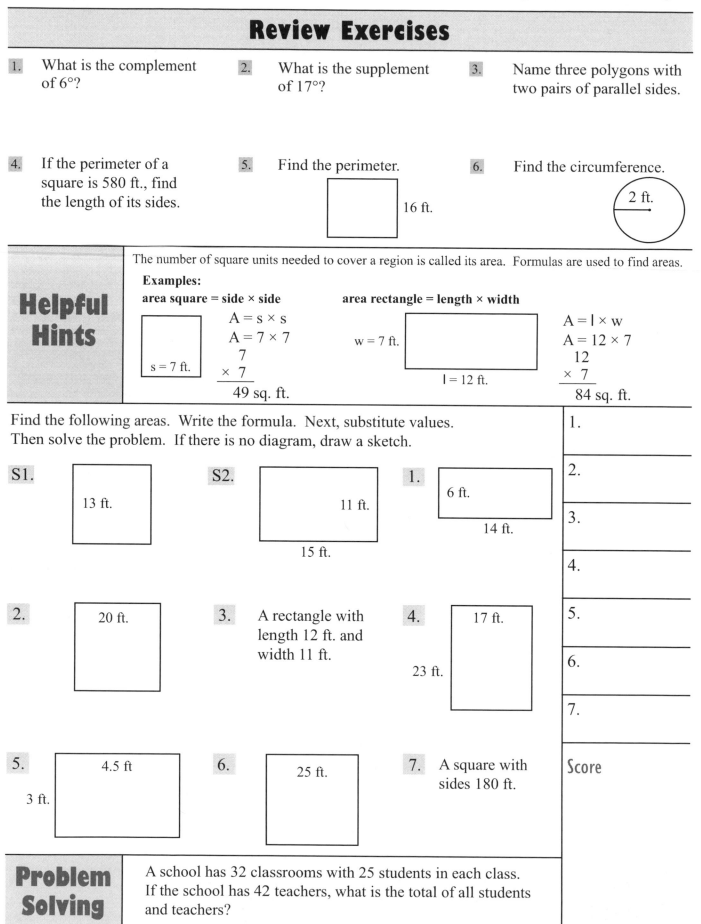

Helpful Hints

The number of square units needed to cover a region is called its area. Formulas are used to find areas.

Examples:

area square = side × side

s = 7 ft.

A = s × s
A = 7 × 7
```
  7
× 7
────
 49 sq. ft.
```

area rectangle = length × width

w = 7 ft.   l = 12 ft.

A = l × w
A = 12 × 7
```
  12
×  7
────
 84 sq. ft.
```

Find the following areas. Write the formula. Next, substitute values. Then solve the problem. If there is no diagram, draw a sketch.

S1.   13 ft.

S2.   11 ft. / 15 ft.

1.   6 ft. / 14 ft.

2.   20 ft.

3.   A rectangle with length 12 ft. and width 11 ft.

4.   17 ft. / 23 ft.

5.   4.5 ft / 3 ft.

6.   25 ft.

7.   A square with sides 180 ft.

1. _____

2. _____

3. _____

4. _____

5. _____

6. _____

7. _____

Score

Problem Solving

A school has 32 classrooms with 25 students in each class. If the school has 42 teachers, what is the total of all students and teachers?

40

Review Exercises

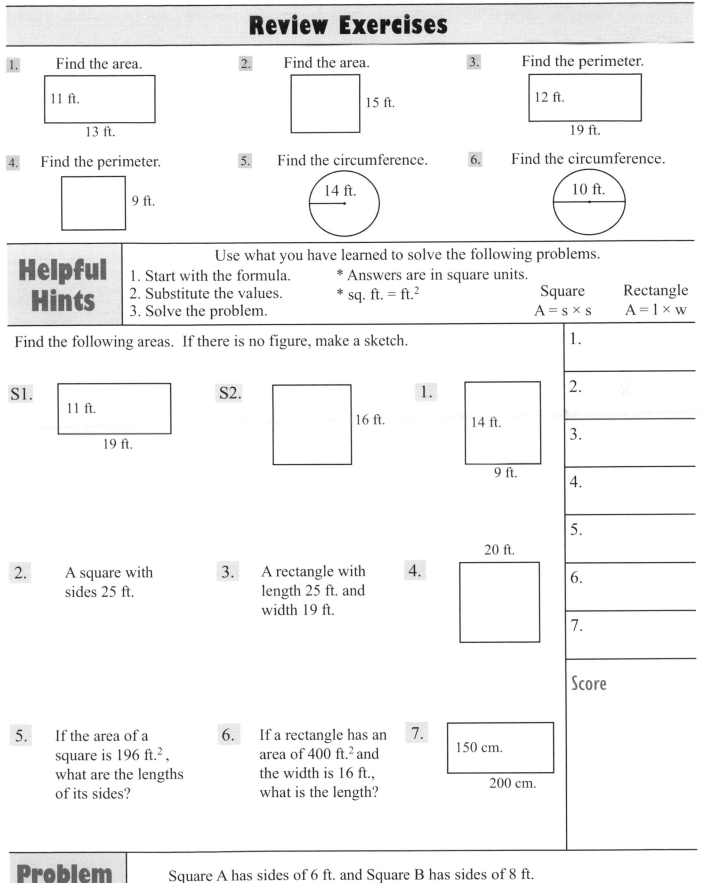

1. Find the area.

    11 ft.

    13 ft.

2. Find the area.

    15 ft.

3. Find the perimeter.

    12 ft.

    19 ft.

4. Find the perimeter.

    9 ft.

5. Find the circumference.

    14 ft.

6. Find the circumference.

    10 ft.

Helpful Hints

Use what you have learned to solve the following problems.
1. Start with the formula.   * Answers are in square units.
2. Substitute the values.   * sq. ft. = ft.^{2}
3. Solve the problem.

	Square	Rectangle
	$A = s \times s$	$A = 1 \times w$

Find the following areas. If there is no figure, make a sketch.

S1.   11 ft.

        19 ft.

S2.   16 ft.

1.   14 ft.

        9 ft.

2. A square with sides 25 ft.

3. A rectangle with length 25 ft. and width 19 ft.

4.   20 ft.

5. If the area of a square is 196 ft.^{2}, what are the lengths of its sides?

6. If a rectangle has an area of 400 ft.^{2} and the width is 16 ft., what is the length?

7.   150 cm.

        200 cm.

1. _____

2. _____

3. _____

4. _____

5. _____

6. _____

7. _____

Score

Problem Solving

Square A has sides of 6 ft. and Square B has sides of 8 ft. How much larger is the area of Square B than Square A?

Review Exercises

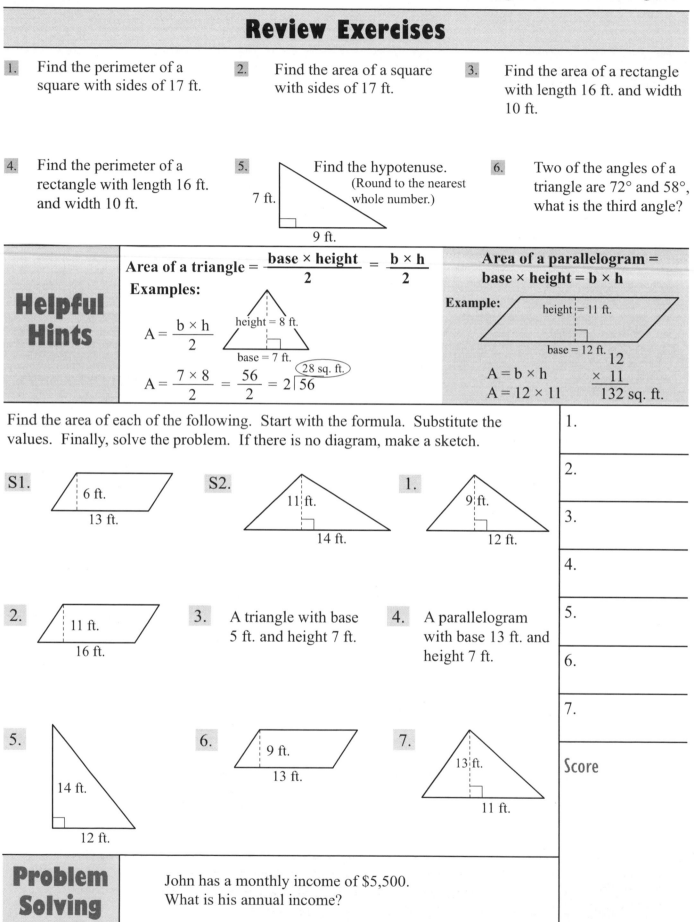

1. Find the perimeter of a square with sides of 17 ft.

2. Find the area of a square with sides of 17 ft.

3. Find the area of a rectangle with length 16 ft. and width 10 ft.

4. Find the perimeter of a rectangle with length 16 ft. and width 10 ft.

5. Find the hypotenuse. (Round to the nearest whole number.)
7 ft. 9 ft.

6. Two of the angles of a triangle are 72° and 58°, what is the third angle?

Helpful Hints

Area of a triangle = $\dfrac{\text{base} \times \text{height}}{2} = \dfrac{b \times h}{2}$

Examples:

$A = \dfrac{b \times h}{2}$

height = 8 ft.
base = 7 ft.

$A = \dfrac{7 \times 8}{2} = \dfrac{56}{2} = 2\overline{)56}$ ⟨28 sq. ft.⟩

Area of a parallelogram = base × height = b × h

Example:
height = 11 ft.
base = 12 ft.

$A = b \times h$
$A = 12 \times 11$

        12
      × 11
    132 sq. ft.

Find the area of each of the following. Start with the formula. Substitute the values. Finally, solve the problem. If there is no diagram, make a sketch.

S1. 6 ft. 13 ft.

S2. 11 ft. 14 ft.

1. 9 ft. 12 ft.

2. 11 ft. 16 ft.

3. A triangle with base 5 ft. and height 7 ft.

4. A parallelogram with base 13 ft. and height 7 ft.

5. 14 ft. 12 ft.

6. 9 ft. 13 ft.

7. 13 ft. 11 ft.

1. _____

2. _____

3. _____

4. _____

5. _____

6. _____

7. _____

Score

Problem Solving

John has a monthly income of $5,500. What is his annual income?

Review Exercises

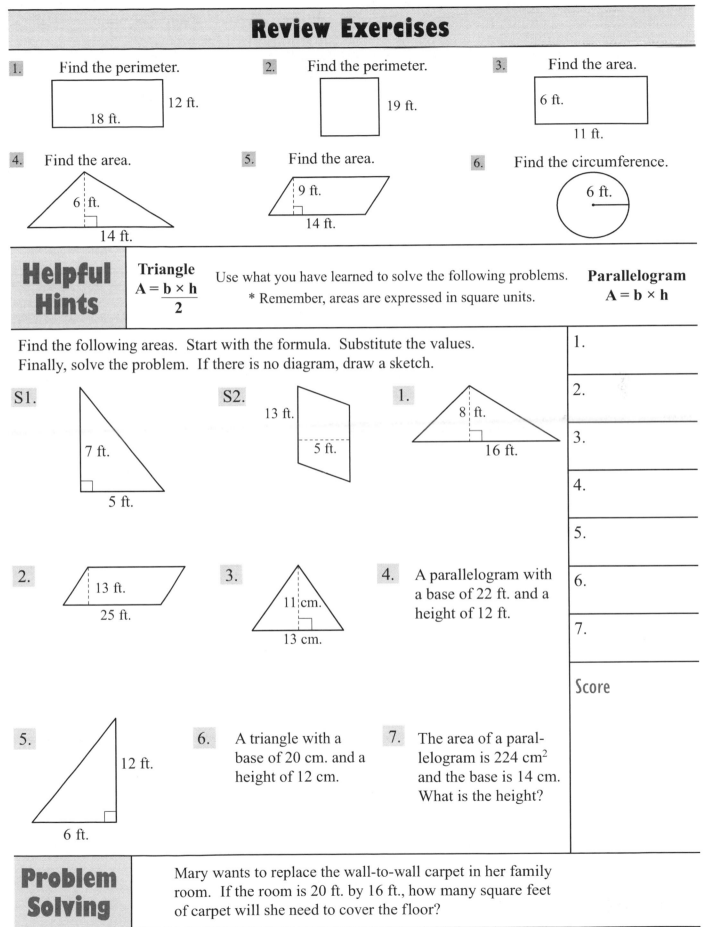

1. Find the perimeter.

    12 ft.
    18 ft.

2. Find the perimeter.

    19 ft.

3. Find the area.

    6 ft.
    11 ft.

4. Find the area.

    6 ft.
    14 ft.

5. Find the area.

    9 ft.
    14 ft.

6. Find the circumference.

    6 ft.

Helpful Hints	**Triangle** $A = \dfrac{b \times h}{2}$	Use what you have learned to solve the following problems. * Remember, areas are expressed in square units.	**Parallelogram** $A = b \times h$

Find the following areas. Start with the formula. Substitute the values. Finally, solve the problem. If there is no diagram, draw a sketch.

S1.

7 ft.
5 ft.

S2.

13 ft.
5 ft.

1.

8 ft.
16 ft.

2.

13 ft.
25 ft.

3.

11 cm.
13 cm.

4. A parallelogram with a base of 22 ft. and a height of 12 ft.

5.

12 ft.
6 ft.

6. A triangle with a base of 20 cm. and a height of 12 cm.

7. The area of a parallelogram is 224 cm^2 and the base is 14 cm. What is the height?

1.

2.

3.

4.

5.

6.

7.

Score

Problem Solving	Mary wants to replace the wall-to-wall carpet in her family room. If the room is 20 ft. by 16 ft., how many square feet of carpet will she need to cover the floor?

Review Exercises

For 1 - 6 find the areas.

1. 14 ft.

2. 8 cm. 12 cm.

3. 9 ft. 12 ft.

4. 7 ft. 15 ft.

5. A triangle with b = 7 ft. and h = 15 ft.

6. A square with sides 8 ft.

Helpful Hints

Area of a trapezoid $= \dfrac{h(B+b)}{2}$

b

h

B

b = 6 ft.

h = 5 ft.

B = 9 ft.

Example: $A = \dfrac{h(B+b)}{2}$

$A = \dfrac{5(9+6)}{2}$

$A = \dfrac{5(15)}{2}$

$A = \dfrac{75}{2} = 2\overline{)75}^{\,37\frac{1}{2}}$ **Area** $= 37\frac{1}{2}$ **sq. ft.**

Find the following areas. If there is no figure, make a sketch.
Write the formula. Substitute the values. Solve the problem.

S1. 3 ft. 5 ft. 7 ft.

S2. 7 ft. 5 ft. 9 ft.

1. 6 ft. 4 ft. 12 ft.

2. A Trapezoid with
B = 16 ft.
b = 6 ft.
h = 4 ft.

3. 7 ft. 6 ft. 14 ft.

4. 10 ft. 12 ft. 14 ft.

5. 8 cm. 14 cm. 6 cm.

6. A Trapezoid with
B = 7 ft.
b = 4 ft.
h = 3 ft.

7. 14 ft. 15 ft. 30 ft.

1. _____
2. _____
3. _____
4. _____
5. _____
6. _____
7. _____

Score

Problem Solving

Zoe is going to make a cloth banner in the shape of a triangle.
If the base of the triangle is 24 inches and the height is 20
inches, how many square inches of cloth does she need?

44

Review Exercises

For 1 - 6 find the perimeter or circumference.

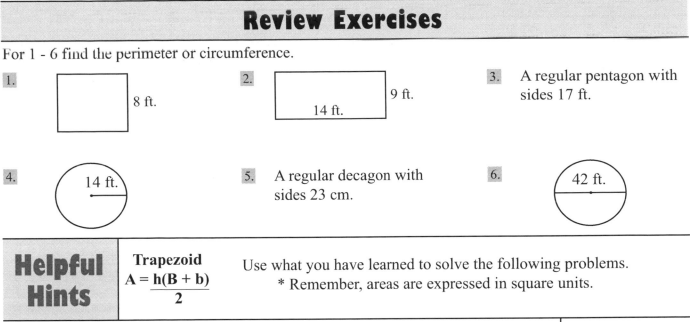

1. 8 ft.

2. 9 ft. / 14 ft.

3. A regular pentagon with sides 17 ft.

4. 14 ft.

5. A regular decagon with sides 23 cm.

6. 42 ft.

Helpful Hints	**Trapezoid** $A = \dfrac{h(B + b)}{2}$	Use what you have learned to solve the following problems. * Remember, areas are expressed in square units.

Find the following areas. If there is no figure, make a sketch.
Write the formula. Substitute the values. Solve the problem.

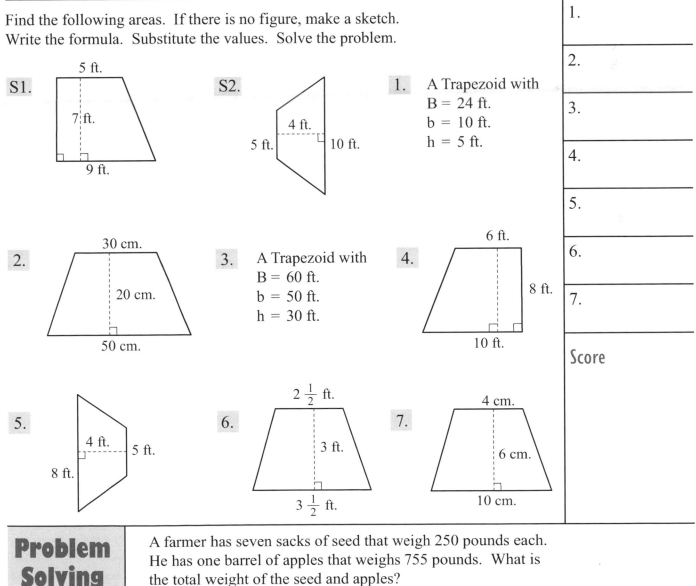

S1. 5 ft. / 7 ft. / 9 ft.

S2. 4 ft. / 5 ft. / 10 ft.

1. A Trapezoid with
   B = 24 ft.
   b = 10 ft.
   h = 5 ft.

2. 30 cm. / 20 cm. / 50 cm.

3. A Trapezoid with
   B = 60 ft.
   b = 50 ft.
   h = 30 ft.

4. 6 ft. / 8 ft. / 10 ft.

5. 4 ft. / 5 ft. / 8 ft.

6. $2\frac{1}{2}$ ft. / 3 ft. / $3\frac{1}{2}$ ft.

7. 4 cm. / 6 cm. / 10 cm.

1.

2.

3.

4.

5.

6.

7.

Score

Problem Solving	A farmer has seven sacks of seed that weigh 250 pounds each. He has one barrel of apples that weighs 755 pounds. What is the total weight of the seed and apples?

Review Exercises

Find the area.
1.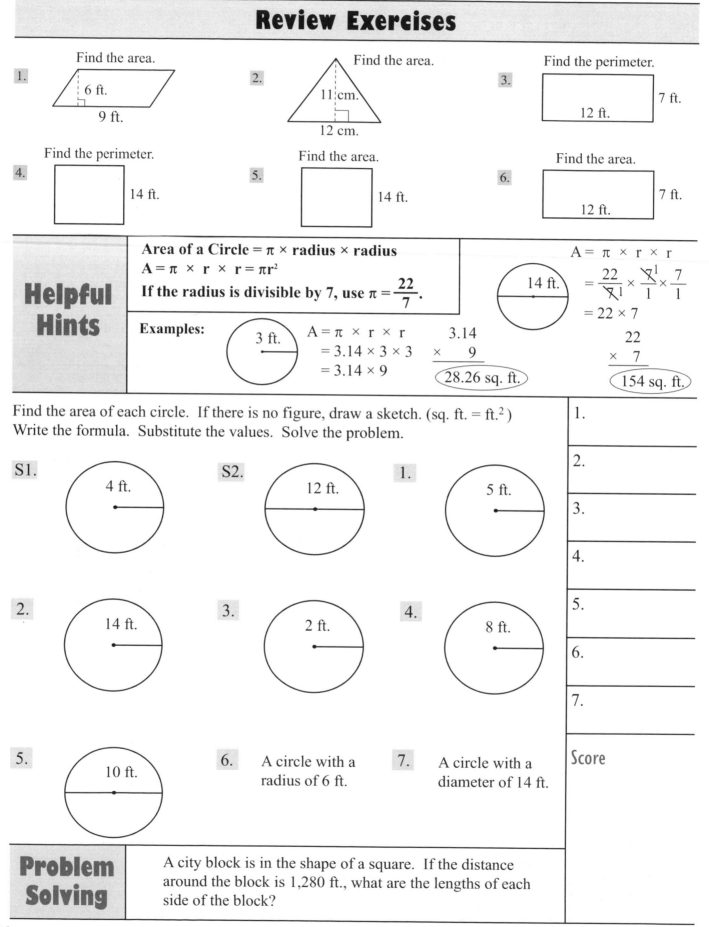
6 ft.
9 ft.

Find the area.
2.
11 cm.
12 cm.

Find the perimeter.
3.
7 ft.
12 ft.

Find the perimeter.
4.
14 ft.

Find the area.
5.
14 ft.

Find the area.
6.
7 ft.
12 ft.

Helpful Hints

Area of a Circle = π × radius × radius
$A = π × r × r = πr^2$
If the radius is divisible by 7, use $π = \frac{22}{7}$.

Examples:
3 ft.
$A = π × r × r$
$= 3.14 × 3 × 3$
$= 3.14 × 9$

3.14
× 9
28.26 sq. ft.

14 ft.
$A = π × r × r$
$= \frac{22}{\cancel{7}^1} × \frac{\cancel{7}^1}{1} × \frac{7}{1}$
$= 22 × 7$

22
× 7
154 sq. ft.

Find the area of each circle. If there is no figure, draw a sketch. (sq. ft. = ft.^{2})
Write the formula. Substitute the values. Solve the problem.

S1.
4 ft.

S2.
12 ft.

1.
5 ft.

2.
14 ft.

3.
2 ft.

4.
8 ft.

5.
10 ft.

6. A circle with a radius of 6 ft.

7. A circle with a diameter of 14 ft.

1.

2.

3.

4.

5.

6.

7.

Score

Problem Solving

A city block is in the shape of a square. If the distance around the block is 1,280 ft., what are the lengths of each side of the block?

Review Exercises

1.   Classify

2. Classify

3. What is the supplement of 70°?

4. Classify
   Sides: _____
   Angles: _____
   60°
   7   7
   60°   60°
   7

5. Find the circumference.
   3 ft.

6. Find the area.
   3 ft.

Helpful Hints

Use what you have learned to solve the following problems.

*Remember areas are expressed in square units.

$$A = \pi \times r \times r$$

*If the radius is divisible by 7, use $\pi = \dfrac{22}{7}$.

Find the area of each circle. If there is no figure, draw a sketch.
Write the formula. Substitute the values. Solve the problem.

S1. 2 km.

S2. 6 ft.

1. 10 ft.

2. 20 ft.

3. A circle with radius 14 ft.

4. 1 ft.

5. A circle with a diameter of 18 ft.

6. 42 ft.

7. 22 ft.

1.
2.
3.
4.
5.
6.
7.

Score

Problem Solving

A school has 280 boys and 320 girls. If the students are grouped into classes of 30 students each, how many classes are there?

Review Exercises

For 1-6, find the area of each figure.

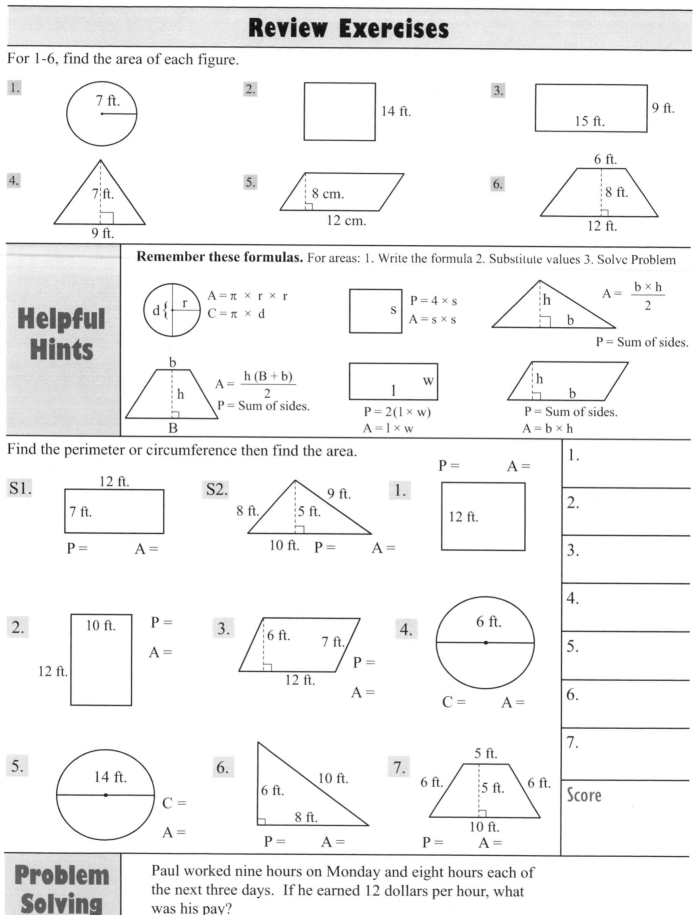

1. 7 ft.

2. 14 ft.

3. 15 ft. 9 ft.

4. 7 ft. 9 ft.

5. 8 cm. 12 cm.

6. 6 ft. 8 ft. 12 ft.

Helpful Hints

Remember these formulas. For areas: 1. Write the formula 2. Substitute values 3. Solve Problem

$A = \pi \times r \times r$
$C = \pi \times d$

$P = 4 \times s$
$A = s \times s$

$A = \dfrac{b \times h}{2}$
P = Sum of sides.

$A = \dfrac{h(B + b)}{2}$
P = Sum of sides.

$P = 2(1 \times w)$
$A = 1 \times w$

P = Sum of sides.
$A = b \times h$

Find the perimeter or circumference then find the area.

S1. 12 ft. 7 ft.
P =   A =

S2. 8 ft. 9 ft. 5 ft. 10 ft.
P =   A =

1. 12 ft.
P =   A =

2. 10 ft. 12 ft.
P =
A =

3. 6 ft. 7 ft. 12 ft.
P =
A =

4. 6 ft.
C =   A =

5. 14 ft.
C =
A =

6. 10 ft. 6 ft. 8 ft.
P =   A =

7. 5 ft. 6 ft. 5 ft. 6 ft. 10 ft.
P =   A =

1.

2.

3.

4.

5.

6.

7.

Score

Problem Solving

Paul worked nine hours on Monday and eight hours each of the next three days. If he earned 12 dollars per hour, what was his pay?

Review Exercises

For 1-6, find the perimeter of each figure.

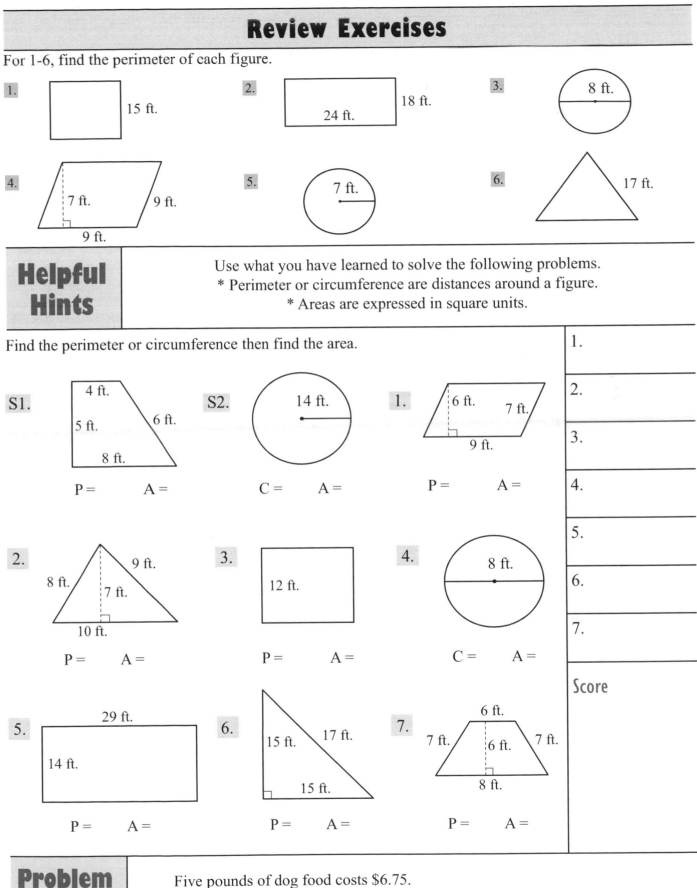

1. 15 ft.

2. 18 ft.
   24 ft.

3. 8 ft.

4. 7 ft.   9 ft.
   9 ft.

5. 7 ft.

6. 17 ft.

Helpful Hints

Use what you have learned to solve the following problems.
* Perimeter or circumference are distances around a figure.
* Areas are expressed in square units.

Find the perimeter or circumference then find the area.

S1. 4 ft.
    5 ft.   6 ft.
    8 ft.
    P =   A =

S2. 14 ft.
    C =   A =

1. 6 ft.   7 ft.
   9 ft.
   P =   A =

2. 9 ft.
   8 ft.   7 ft.
   10 ft.
   P =   A =

3. 12 ft.
   P =   A =

4. 8 ft.
   C =   A =

5. 29 ft.
   14 ft.
   P =   A =

6. 15 ft.   17 ft.
   15 ft.
   P =   A =

7. 6 ft.
   7 ft.   6 ft.   7 ft.
   8 ft.
   P =   A =

1.

2.

3.

4.

5.

6.

7.

Score

Problem Solving

Five pounds of dog food costs $6.75.
What is the price per pound?

Review Exercises

For 1 - 6 draw a sketch and find the area.

1. **Square:**
Sides, 16 ft.

2. **Triangle:**
Base, 18 ft.
Height, 8 ft.

3. **Parallelogram:**
Base, 22 ft.
Height, 16 ft.

4. **Trapezoid:**
B, 22 ft.
b, 20 ft.
h, 6 ft.

5. **Rectangle:**
Length, 16 ft.
Width, 18 ft.

6. **Circle:**
Radius, 5 ft.

Helpful Hints

Use what you have learned to solve the following problems.
1. Read the problem carefully to understand what is being asked.
2. Draw a sketch.
3. Write the formula. Substitute the values. Solve the problem.
   Label the answer with a word or short phrase.

S1. The Smith's kitchen floor is 12 ft. by 14 ft. What is the perimeter of the floor?

S2. Phil's garden is 12 ft. by 10 ft. If each bag of fertilizer will cover 6 sq. ft., how many bags of fertilizer will he need to care of his garden?

1. Anna wants to build a fence around her yard. If the rectangular yard is 28 ft. by 20 ft., how many feet long will the fence be?

2. Eddie wants to put crown molding around the ceiling of his living room. The room is rectangular and is 24 ft. by 18 ft. If molding comes in 6 foot sections, how many sections of molding must he buy?

3. Find the circumference of a circular flower bed with radius 8 ft.

4. Sue wants to paint a rectangular wall that is 15 ft. by 9 ft. A can of paint covers 45 sq. ft. How many cans must she buy?

5. The perimeter of a square is 624 ft. What is the length of its sides?

6. A city is in the shape of a rectangle. Its area is 192 sq. miles and its width is 12 miles. What is the city's length?

7. A walking path is in the shape of an equilateral triangle. If Dan has walked the first two sides and covered 1260 meters, what is the length of each side of the walking path?

| 1. |
| 2. |
| 3. |
| 4. |
| 5. |
| 6. |
| 7. |
| Score |

Problem Solving

Mr. Jenkins earned $3,600 and deposited $\frac{1}{3}$ of it into his savings account. How much did he deposit into his savings account?

Review Exercises

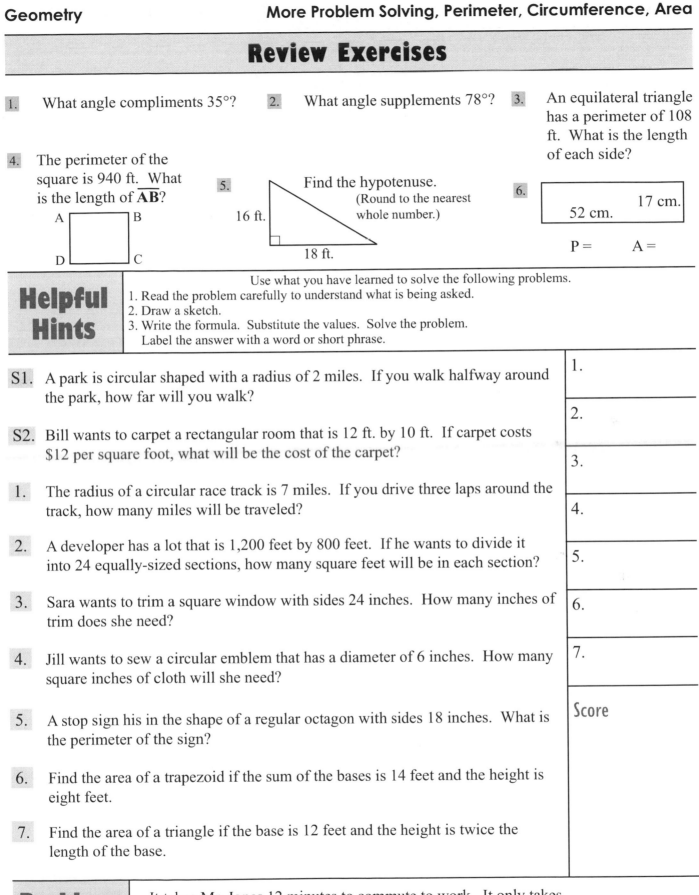

1. What angle compliments 35°?

2. What angle supplements 78°?

3. An equilateral triangle has a perimeter of 108 ft. What is the length of each side?

4. The perimeter of the square is 940 ft. What is the length of $\overline{AB}$?

A   B

D   C

5. Find the hypotenuse. (Round to the nearest whole number.)

16 ft.   18 ft.

6.

   17 cm.
52 cm.

P =   A =

Helpful Hints

Use what you have learned to solve the following problems.
1. Read the problem carefully to understand what is being asked.
2. Draw a sketch.
3. Write the formula. Substitute the values. Solve the problem.
   Label the answer with a word or short phrase.

S1. A park is circular shaped with a radius of 2 miles. If you walk halfway around the park, how far will you walk?

S2. Bill wants to carpet a rectangular room that is 12 ft. by 10 ft. If carpet costs $12 per square foot, what will be the cost of the carpet?

1. The radius of a circular race track is 7 miles. If you drive three laps around the track, how many miles will be traveled?

2. A developer has a lot that is 1,200 feet by 800 feet. If he wants to divide it into 24 equally-sized sections, how many square feet will be in each section?

3. Sara wants to trim a square window with sides 24 inches. How many inches of trim does she need?

4. Jill wants to sew a circular emblem that has a diameter of 6 inches. How many square inches of cloth will she need?

5. A stop sign his in the shape of a regular octagon with sides 18 inches. What is the perimeter of the sign?

6. Find the area of a trapezoid if the sum of the bases is 14 feet and the height is eight feet.

7. Find the area of a triangle if the base is 12 feet and the height is twice the length of the base.

1.

2.

3.

4.

5.

6.

7.

Score

Problem Solving

It takes Mr. Jones 12 minutes to commute to work. It only takes 10 minutes to drive home. How many minutes does it take him to drive to and from work in a 5-day work week?

Review Exercises

1. Find the circumference.

4 ft.

2. Find the area.

7 ft.

16 ft.

3. Find the perimeter.

95 cm.

4. Find the area.

14 ft.

3 ft.

18 ft.

5. Find the circumference.

21 ft.

6. Find the area.

11 ft.

15 ft.

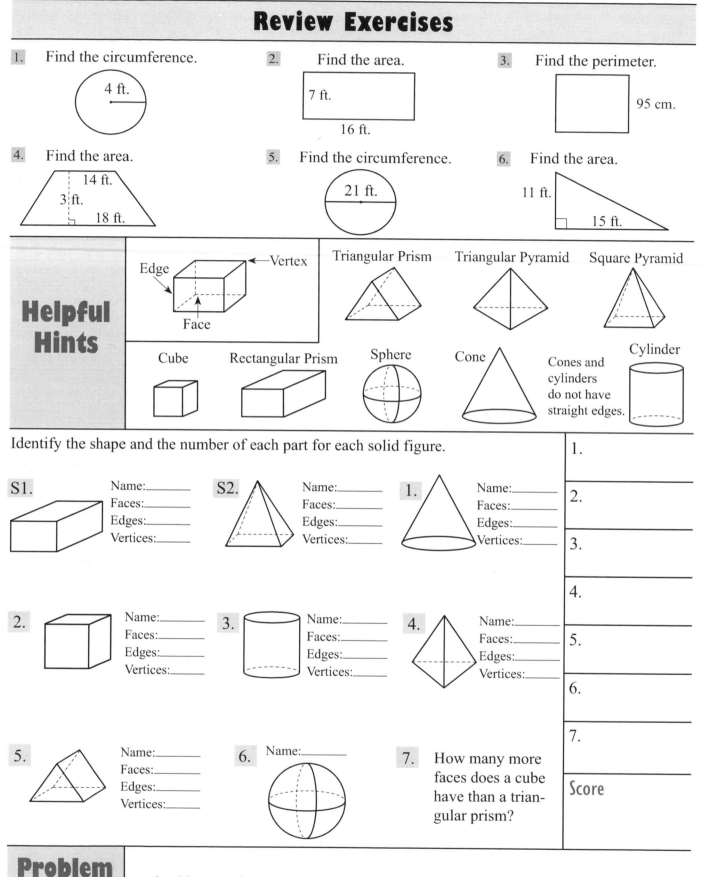

Helpful Hints

Edge   Vertex

Face

Triangular Prism   Triangular Pyramid   Square Pyramid

Cube   Rectangular Prism   Sphere   Cone   Cones and cylinders do not have straight edges.   Cylinder

Identify the shape and the number of each part for each solid figure.

S1. Name:_____ Faces:_____ Edges:_____ Vertices:_____

S2. Name:_____ Faces:_____ Edges:_____ Vertices:_____

1. Name:_____ Faces:_____ Edges:_____ Vertices:_____

2. Name:_____ Faces:_____ Edges:_____ Vertices:_____

3. Name:_____ Faces:_____ Edges:_____ Vertices:_____

4. Name:_____ Faces:_____ Edges:_____ Vertices:_____

5. Name:_____ Faces:_____ Edges:_____ Vertices:_____

6. Name:_____

7. How many more faces does a cube have than a triangular prism?

1.

2.

3.

4.

5.

6.

7.

Score

Problem Solving

Cookies cost $.32 each. How much will two dozen cost?

Review Exercises

1. Name a solid shape with five faces.

2. Name a solid shape with six faces.

3. Which solid shape has no faces?

4. Name two solid shapes with 12 edges.

5. Name a solid shape with five vertices.

6. Name the solid shape with one or two faces.

Helpful Hints

Use what you have learned about solid shapes to answer the following questions.

For each solid shape list four real life objects that have that shape.

S1. Sphere

S2. Triangular Prism

1. Cube

2. Cone

3. Cylinder

4. Rectangular Prism

5. Square Pyramid

6. Triangular Pyramid

7. Objects made up of a combination of solid shapes.

1.

2.

3.

4.

5.

6.

7.

8.

9.

10.

Score

Problem Solving

If a city has an average monthly rainfall of 15 inches, what is its average annual rainfall?

Review Exercises

1. Find the complement of 18°.

2. Find the supplement of 114°.

3. Find the area of a circle with a radius of five feet.

4. Find the circumference of a circle with a radius of five feet.

5. Find the perimeter of a regular hexagon with sides of 15 ft.

6. Find the area of a square with sides of 16 ft.

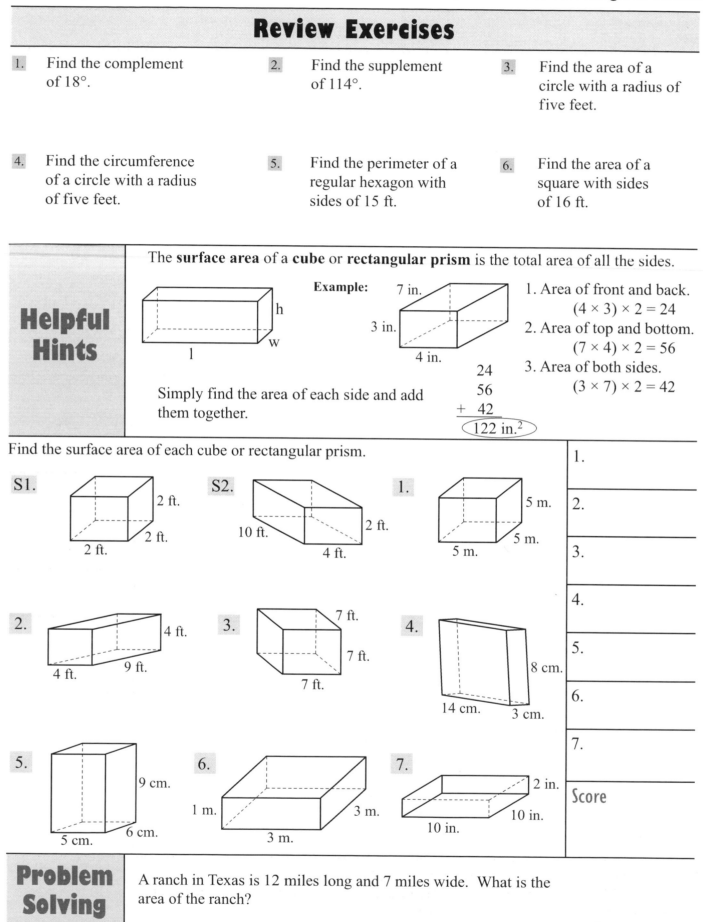

Helpful Hints

The **surface area** of a **cube** or **rectangular prism** is the total area of all the sides.

Simply find the area of each side and add them together.

Example: 7 in.   3 in.   4 in.

1. Area of front and back.
   (4 × 3) × 2 = 24
2. Area of top and bottom.
   (7 × 4) × 2 = 56
3. Area of both sides.
   (3 × 7) × 2 = 42

   24
   56
 + 42
 122 in.²

Find the surface area of each cube or rectangular prism.

S1.   2 ft.   2 ft.   2 ft.

S2.   10 ft.   2 ft.   4 ft.

1.   5 m.   5 m.   5 m.

2.   4 ft.   9 ft.   4 ft.

3.   7 ft.   7 ft.   7 ft.

4.   8 cm.   14 cm.   3 cm.

5.   9 cm.   5 cm.   6 cm.

6.   1 m.   3 m.   3 m.

7.   2 in.   10 in.   10 in.

1. _____
2. _____
3. _____
4. _____
5. _____
6. _____
7. _____

Score

Problem Solving

A ranch in Texas is 12 miles long and 7 miles wide. What is the area of the ranch?

Review Exercises

For 1 - 6 find the area of each figure.

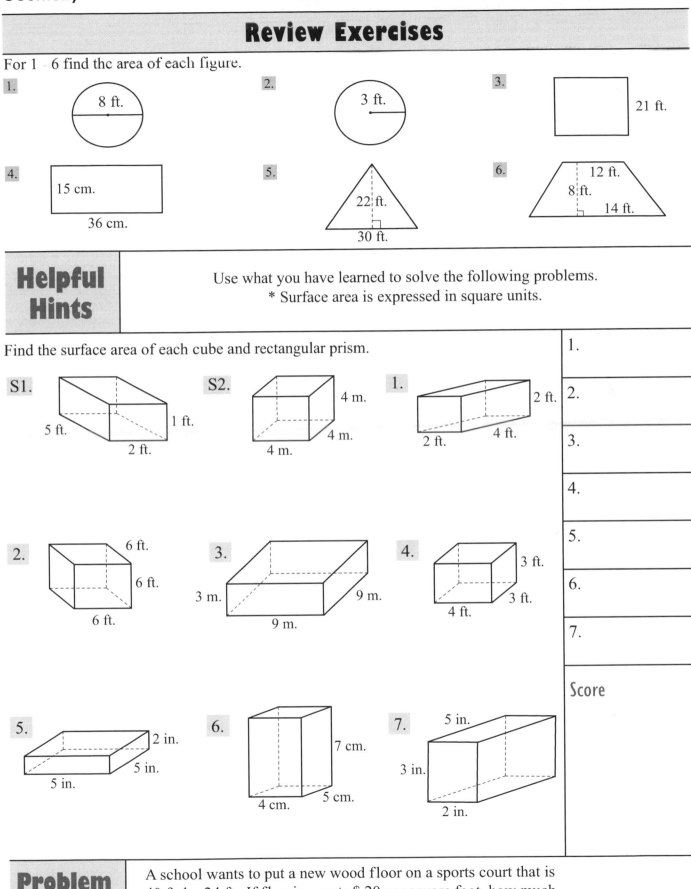

Helpful Hints

Use what you have learned to solve the following problems.
* Surface area is expressed in square units.

Find the surface area of each cube and rectangular prism.

1.
2.
3.
4.
5.
6.
7.
Score

Problem Solving

A school wants to put a new wood floor on a sports court that is 40 ft. by 24 ft. If flooring costs $ 20 per square foot, how much will the new floor cost?

Review Exercises

For 1 - 6 find the perimeter or circumference.

1. A square with sides of 29 ft.

2. A rectangle with length 16 ft. and width 7 ft.

3. A circle with a diameter of 14 ft.

4. A circle with a radius of 3 ft.

5. A regular pentagon with sides of 14 ft.

6. An equilateral triangle with sides of 38 ft.

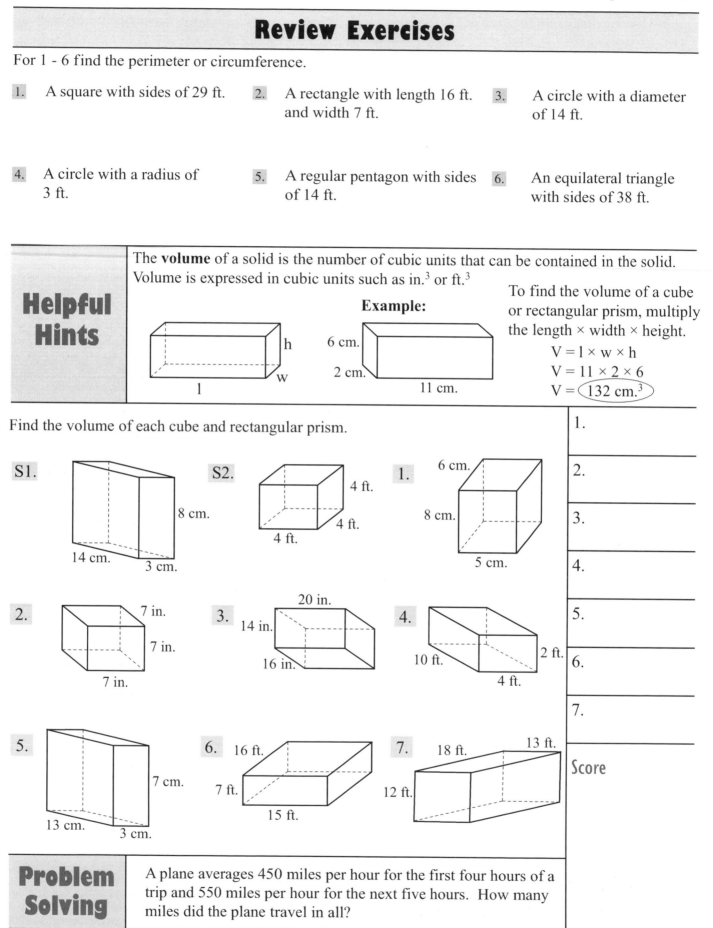

Helpful Hints

The **volume** of a solid is the number of cubic units that can be contained in the solid. Volume is expressed in cubic units such as in.³ or ft.³

Example:

To find the volume of a cube or rectangular prism, multiply the length × width × height.

$V = l \times w \times h$

$V = 11 \times 2 \times 6$

$V = 132 \text{ cm.}^3$

Find the volume of each cube and rectangular prism.

S1. 8 cm. 14 cm. 3 cm.

S2. 4 ft. 4 ft. 4 ft.

1. 6 cm. 8 cm. 5 cm.

2. 7 in. 7 in. 7 in.

3. 20 in. 14 in. 16 in.

4. 10 ft. 4 ft. 2 ft.

5. 7 cm. 13 cm. 3 cm.

6. 16 ft. 7 ft. 15 ft.

7. 18 ft. 13 ft. 12 ft.

1.

2.

3.

4.

5.

6.

7.

Score

Problem Solving

A plane averages 450 miles per hour for the first four hours of a trip and 550 miles per hour for the next five hours. How many miles did the plane travel in all?

56

Review Exercises

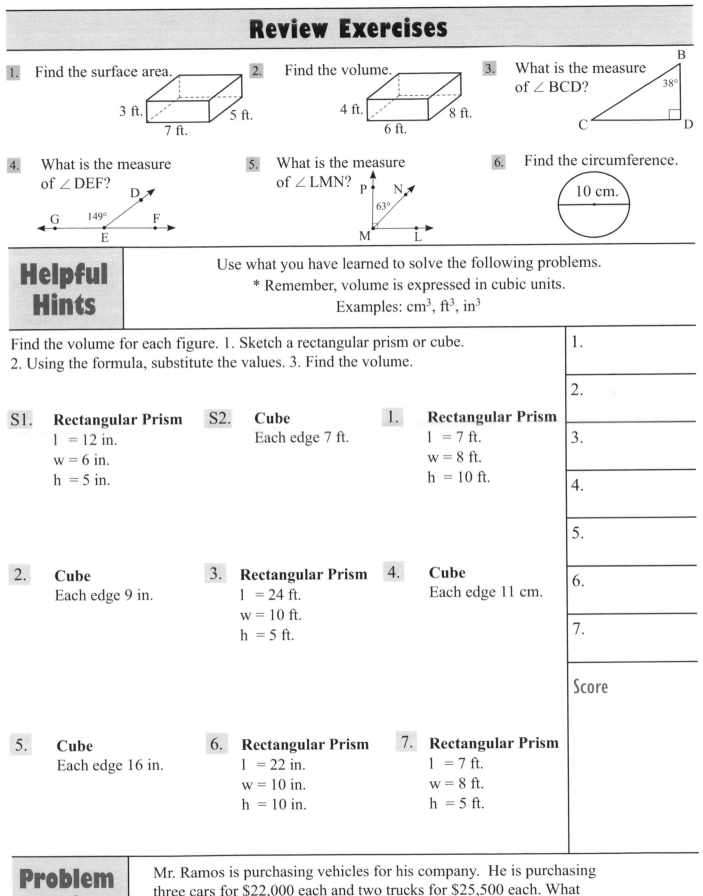

1. Find the surface area.

3 ft. 5 ft. 7 ft.

2. Find the volume.

4 ft. 8 ft. 6 ft.

3. What is the measure of ∠ BCD?

B
38°
C   D

4. What is the measure of ∠ DEF?

D
G 149° F
E

5. What is the measure of ∠ LMN?

P N
63°
M L

6. Find the circumference.

10 cm.

Helpful Hints

Use what you have learned to solve the following problems.
* Remember, volume is expressed in cubic units.
Examples: cm³, ft³, in³

Find the volume for each figure. 1. Sketch a rectangular prism or cube.
2. Using the formula, substitute the values. 3. Find the volume.

S1. **Rectangular Prism**
l = 12 in.
w = 6 in.
h = 5 in.

S2. **Cube**
Each edge 7 ft.

1. **Rectangular Prism**
l = 7 ft.
w = 8 ft.
h = 10 ft.

2. **Cube**
Each edge 9 in.

3. **Rectangular Prism**
l = 24 ft.
w = 10 ft.
h = 5 ft.

4. **Cube**
Each edge 11 cm.

5. **Cube**
Each edge 16 in.

6. **Rectangular Prism**
l = 22 in.
w = 10 in.
h = 10 in.

7. **Rectangular Prism**
l = 7 ft.
w = 8 ft.
h = 5 ft.

1.

2.

3.

4.

5.

6.

7.

Score

Problem Solving

Mr. Ramos is purchasing vehicles for his company. He is purchasing three cars for $22,000 each and two trucks for $25,500 each. What will be the total cost of the vehicles?

Review Exercises

1. Find the surface area.

3 ft. / 10 ft. / 7 ft.

2. Find the volume.

4 ft. / 3 ft. / 7 ft.

3. Find the surface area of the cube.

4 ft.

4. Find the volume of the cube.

4 ft.

5. Find the perimeter.

57 cm.

6. Find the area.

19 ft.

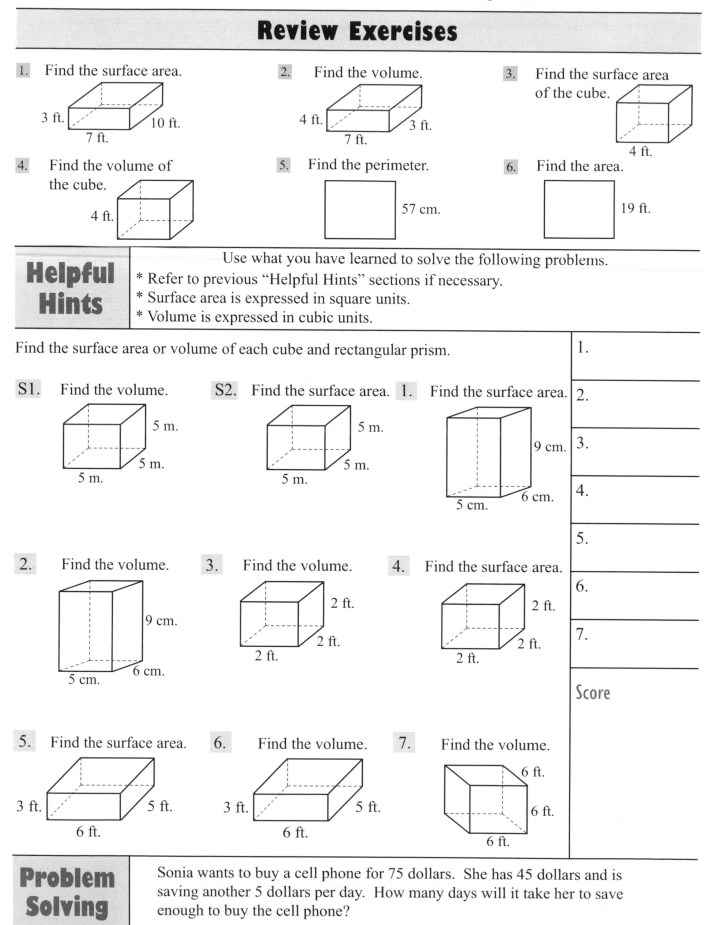

Helpful Hints

Use what you have learned to solve the following problems.
* Refer to previous "Helpful Hints" sections if necessary.
* Surface area is expressed in square units.
* Volume is expressed in cubic units.

Find the surface area or volume of each cube and rectangular prism.

S1. Find the volume.

5 m. / 5 m. / 5 m.

S2. Find the surface area.

5 m. / 5 m. / 5 m.

1. Find the surface area.

9 cm. / 5 cm. / 6 cm.

2. Find the volume.

9 cm. / 5 cm. / 6 cm.

3. Find the volume.

2 ft. / 2 ft. / 2 ft.

4. Find the surface area.

2 ft. / 2 ft. / 2 ft.

5. Find the surface area.

3 ft. / 5 ft. / 6 ft.

6. Find the volume.

3 ft. / 5 ft. / 6 ft.

7. Find the volume.

6 ft. / 6 ft. / 6 ft.

1.

2.

3.

4.

5.

6.

7.

Score

Problem Solving

Sonia wants to buy a cell phone for 75 dollars. She has 45 dollars and is saving another 5 dollars per day. How many days will it take her to save enough to buy the cell phone?

Review Exercises

For 1 - 6 find the area of each figure.

1. [trapezoid: 8 ft. top, 6 ft. height, 16 ft. bottom]

2. [parallelogram: 18 cm., 19 cm.]

3. [square: 16 in.]

4. [rectangle: 9 in., 15 in.]

5. [triangle: 9 ft., 9 ft.]

6. [circle: 4 ft.]

Helpful Hints	Use what you have learned to solve the following problems.
	* Refer to previous "Helpful Hints" sections if necessary.
	* Surface area is expressed in square units.
	* Volume is expressed in cubic units.

Solve each of the following. Make a sketch. Next, substitute values.
Finally, solve the problem.

S1. Find the volume of a rectangular prism with a length of 12 ft., width of 9 ft., and height of 7 ft.

S2. Find the surface area of a cube with edges of 4 inches in length.

1. A shed with a flat roof is 10 feet long, 8 feet wide, and 7 feet tall. What is the surface area of the shed?

2. A gift box in the shape of a cube has edges of 9 inches. How many square inches of wrapping paper is needed to cover the box?

3. An aquarium is 15 inches long, 10 inches wide, and 10 inches high. What is the volume of the aquarium?

4. Gill is going to paint the outside of a box which is six feet long, four feet wide, and five feet high. How many square feet of surface area will he paint?

5. The volume of a rectangular prism is 2,184 in.^{3}. If the length is 12 in. and the height is 14 in., what is the width?

1.

2.

3.

4.

5.

Score

Problem Solving   Nathan worked 1,840 hours last year. If his hourly pay was 18 dollars per hour, how much did he earn last year?

59

Final Review

Use the figure to answer problems 1 - 10.

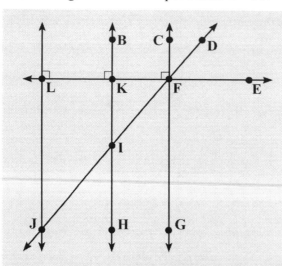

1. Name two parallel lines.
2. Name two perpendicular lines.
3. Name three line segments.
4. Name three rays.
5. Name two acute angles.
6. Name one triangle.
7. Name two obtuse angles.
8. Name one straight angle.
9. Name two right angles.
10. Name one trapezoid.

Use the figures below to answer problems 11 - 12.

Triangle A   Triangle B

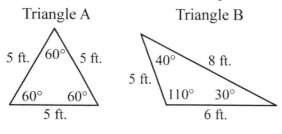

11. Classify triangle A by its sides and angles.
12. Classify triangle B by its sides and angles.

For problems 13 - 17 use the circle below.

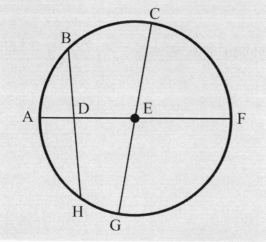

13. Name two diameters.
14. Name three radii.
15. Name two chords.
16. If $\overline{CG}$ is 24 ft., what is the length of $\overline{CE}$?
17. What part of the circle is $\overline{BH}$?

For problems 18 - 20 identify the shape and numbers of each part.

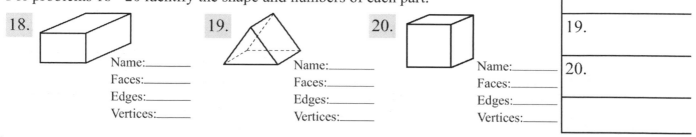

18.
Name:_____
Faces:_____
Edges:_____
Vertices:_____

19.
Name:_____
Faces:_____
Edges:_____
Vertices:_____

20.
Name:_____
Faces:_____
Edges:_____
Vertices:_____

1.
2.
3.
4.
5.
6.
7.
8.
9.
10.
11.
12.
13.
14.
15.
16.
17.
18.
19.
20.

Final Test Concepts

Use the figure to answer problems 1 - 10.

1. | 1.
2. | 2.
3. |

1. Name two parallel lines.
2. Name two perpendicular lines.
3. Name an obtuse, an acute and a right angle.
4. Name two rays.
5. Name two triangles.
6. Name a trapezoid.
7. Name two line segments.
8. Name one pentagon.
9. Name one straight angle.
10. Name two congruent line segments.

Use the figures below to answer problems 11 - 13.

Triangle A   Triangle B

5 ft. 85° 7 ft.   C 60° 5 ft.
60°   35°   A   30°
    9 ft.   4 ft. B

11. Classify triangle A by its sides and angles.
12. Classify triangle B by its sides and angles.
13. In triangle B, what part of the triangle is $\overline{CB}$?

14. What polygon has one pair of parallel sides?

15. Name three polygons that have two pairs of parallel sides.

16. What polygon has eight sides?

17. What term is given to a polygon with all its angles congruent and all its sides congruent?

For problems 18 - 20 identify the shape and numbers of each part.

18.   19.   20.

Name:_____   Name:_____   Name:_____
Faces:_____   Faces:_____   Faces:_____
Edges:_____   Edges:_____   Edges:_____
Vertices:_____   Vertices:_____   Vertices:_____

1.
2.
3.
4.
5.
6.
7.
8.
9.
10.
11.
12.
13.
14.
15.
16.
17.
18.
19.
20.

Final Review

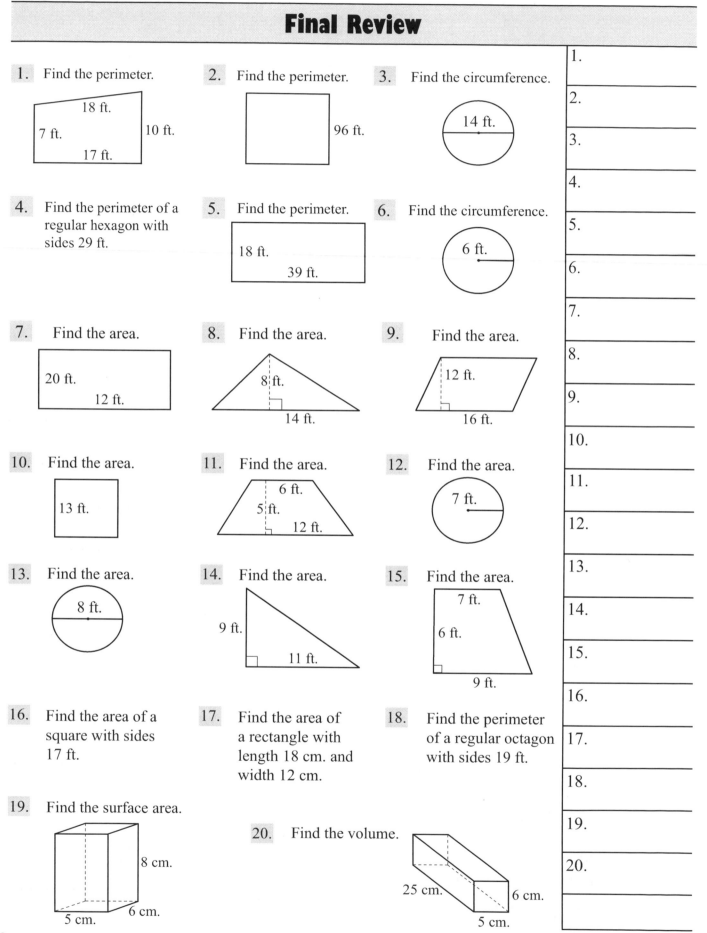

1. Find the perimeter.

18 ft.
7 ft.   10 ft.
17 ft.

2. Find the perimeter.

96 ft.

3. Find the circumference.

14 ft.

4. Find the perimeter of a regular hexagon with sides 29 ft.

5. Find the perimeter.

18 ft.
39 ft.

6. Find the circumference.

6 ft.

7. Find the area.

20 ft.
12 ft.

8. Find the area.

8 ft.
14 ft.

9. Find the area.

12 ft.
16 ft.

10. Find the area.

13 ft.

11. Find the area.

6 ft.
5 ft.
12 ft.

12. Find the area.

7 ft.

13. Find the area.

8 ft.

14. Find the area.

9 ft.
11 ft.

15. Find the area.

7 ft.
6 ft.
9 ft.

16. Find the area of a square with sides 17 ft.

17. Find the area of a rectangle with length 18 cm. and width 12 cm.

18. Find the perimeter of a regular octagon with sides 19 ft.

19. Find the surface area.

8 cm.
5 cm.   6 cm.

20. Find the volume.

25 cm.   6 cm.
5 cm.

1.
2.
3.
4.
5.
6.
7.
8.
9.
10.
11.
12.
13.
14.
15.
16.
17.
18.
19.
20.

62

Final Test

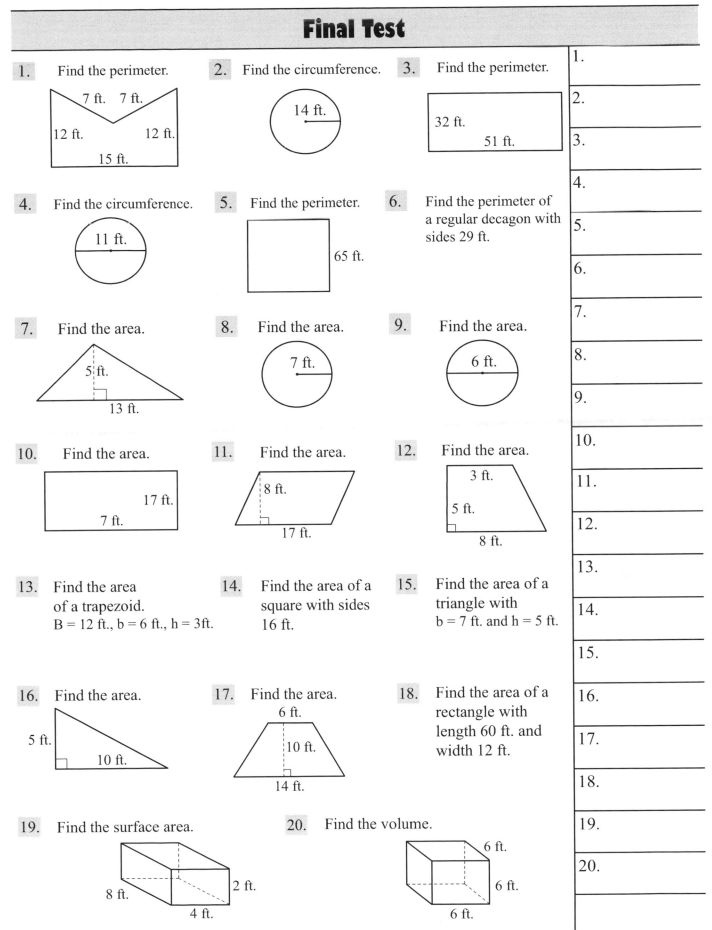

1. Find the perimeter.

    7 ft.   7 ft.
    12 ft.   12 ft.
          15 ft.

2. Find the circumference.

    14 ft.

3. Find the perimeter.

    32 ft.
          51 ft.

4. Find the circumference.

    11 ft.

5. Find the perimeter.

    65 ft.

6. Find the perimeter of a regular decagon with sides 29 ft.

7. Find the area.

    5 ft.
        13 ft.

8. Find the area.

    7 ft.

9. Find the area.

    6 ft.

10. Find the area.

    17 ft.
    7 ft.

11. Find the area.

    8 ft.
        17 ft.

12. Find the area.

    3 ft.
    5 ft.
        8 ft.

13. Find the area of a trapezoid.
    B = 12 ft., b = 6 ft., h = 3ft.

14. Find the area of a square with sides 16 ft.

15. Find the area of a triangle with b = 7 ft. and h = 5 ft.

16. Find the area.

    5 ft.
        10 ft.

17. Find the area.
    6 ft.
    10 ft.
        14 ft.

18. Find the area of a rectangle with length 60 ft. and width 12 ft.

19. Find the surface area.

    8 ft.   2 ft.
        4 ft.

20. Find the volume.

    6 ft.
    6 ft.
    6 ft.

1.
2.
3.
4.
5.
6.
7.
8.
9.
10.
11.
12.
13.
14.
15.
16.
17.
18.
19.
20.

Solutions

Page 4

Review Exercises
1. 1,085
2. 636
3. 1,482
4. 184
5. 584
6. 3,152

S1. answers vary
S2. answers vary
1. answers vary
2. answers vary
3. D, E, F
4. $\overleftrightarrow{AC}$, $\overleftrightarrow{BC}$
5. $\overrightarrow{FE}$, $\overrightarrow{FB}$
6. $\overrightarrow{AB}$, $\overrightarrow{BC}$
7. $\overline{FE}$, $\overline{ED}$, $\overline{FD}$
8. answers vary
9. answers vary
10. Point E

Problem Solving: $6,252

Page 5

Review Exercises
1. 1,087
2. 109
3. 14,959
4. 1,058
5. 1,180
6. 4,213

S1. answers vary
S2. answers vary
1. answers vary
2. answers vary
3. G, I, D, C
4. answers vary
5. $\overrightarrow{JH}$, $\overrightarrow{IH}$
6. $\overrightarrow{EA}$, $\overrightarrow{DA}$
7. answers vary
8. answers vary
9. answers vary
10. D

Problem Solving: 18,200 cars

Page 6

Review Exercises
1. 212
2. 123 r2
3. 296 r4
4. 4,606
5. 1,397
6. 4,343

S1. $\overleftrightarrow{DK}$, $\overleftrightarrow{EJ}$
S2. answers vary
1. answers vary
2. answers vary
3. answers vary
4. answers vary
5. answers vary
6. answers vary
7. answers vary
8. answers vary
9. answers vary
10. ∠H, ∠IHJ

Problem Solving: 87

Page 7

Review Exercises
1. answers vary
2. answers vary
3. answers vary
4. 7,248
5. 6,445
6. 121

S1. answers vary
S2. $\overleftrightarrow{BM}$, $\overleftrightarrow{CF}$
1. answers vary
2. answers vary
3. answers vary
4. answers vary
5. answers vary
6. answers vary
7. answers vary
8. answers vary
9. answers vary
10. ∠L, ∠HLB

Problem Solving: $154

Page 8

Review Exercises
1. 1,147
2. answers vary
3. 735
4. answers vary
5. 3,380
6. 17,220

S1. answers vary
S2. answers vary
1. answers vary
2. answers vary
3. acute
4. obtuse
5. right
6. straight
7. answers vary
8. answers vary
9. answers vary
10. answers vary

Problem Solving:
55 miles per hour

Page 9

Review Exercises
1. 1,631
2. 4,249
3. 648
4. 19,200
5. 146,730
6. 1,001

S1. answers vary
S2. answers vary
1. answers vary
2. answers vary
3. acute
4. obtuse
5. right
6. 60°
7. straight
8. acute
9. obtuse
10. 150°

Problem Solving:
27 are empty

Page 10

Review Exercises
1. acute
2. obtuse
3. answers vary
4. answers vary
5. answers vary
6. 8,234

S1. acute, 20°
S2. obtuse, 110°
1. right, 90°
2. obtuse, 160°
3. acute, 20°
4. acute, 70°
5. obtuse, 130°
6. acute, 50°
7. straight, 180°
8. obtuse, 160°
9. right, 90°
10. obtuse, 130°

Problem Solving: 21 students

Page 11

Review Exercises
1. answers vary
2. answers vary
3. answers vary
4. answers vary
5. 1,691
6. 4,428

S1. 20°, acute
S2. 65°, acute
1. 100°, obtuse
2. 145°, obtuse
3. 45°, acute
4. 35°, acute
5. 180°, straight
6. 80°, acute
7. 60°, acute
8. 160°, obtuse
9. 95°, acute
10. 125°, obtuse

Problem Solving: 87 boxes

Page 12

Review Exercises
1. 205
2. 221
3. 1,155
4. 22,176
5. 2,079
6. 1,118

Answers for S1 - 10
are approximate.

S1. 43°, acute
S2. 104°, obtuse
1. 60°, acute
2. 74°, acute
3. 106°, obtuse
4. 120°, obtuse
5. 77°, acute
6. 137°, obtuse
7. 160°, obtuse
8. 36°, acute
9. 113°, obtuse
10. 20°, acute

Problem Solving: 12 gallons

Page 13

Review Exercises
1. answers vary
2. answers vary
3. answers vary
4. answers vary
5. 12°
6. 28°

Answers for S1 - 10
are approximate.

S1. 23°, acute
S2. 104°, obtuse
1. 60°, acute
2. 127°, obtuse
3. 77°, acute
4. 45°, acute
5. 118°, obtuse
6. 104°, obtuse
7. 77°, acute
8. 144°, obtuse
9. 54°, acute
10. 11°, acute

Problem Solving:
    8 gallons, $24

Page 14

Review Exercises
1. 37°
2. 6,776
3. answers vary
4. answers vary
5. answers vary
6. answers vary

S1. 39°
S2. 93°
1. 18°
2. 74°
3. 85°
4. 38°
5. 19°
6. 105°
7. 168°
8. 165°
9. 8°
10. 67°

Problem Solving: 21 miles

Page 15

Review Exercises
1. 47°
2. 73°
3. 1,242
4. answers vary
5. 1,748
6. 3,969

S1. 115°
S2. 31°
1. 101°
2. 39°
3. 147°
4. 161°
5. 118°
6. 78°
7. 47°
8. 79°
9. 34°
10. 42°

Problem Solving: 8,000 sheets

Page 16

Review Exercises
1. 76°
2. 166°
3. 32°
4. 28°
5. 266
6. 6,755

S1. answers vary
S2. answers vary
1. answers vary
2. answers vary
3. answers vary
4. answers vary
5. answers vary
6. answers vary
7. answers vary

Problem Solving: $54,000

Page 17

Review Exercises
1. 98°
2. 73°
3. 5,505
4. 39°
5. 142°
6. answers vary

S1. answers vary
S2. answers vary
1. answers vary
2. answers vary
3. answers vary
4. answers vary
5. answers vary
6. answers vary
7. answers vary

Problem Solving: $46

Page 18

Review Exercises
1. ∠ DEF, acute
2. ∠ HGF, obtuse
3. ∠ JKL, right
4. parallel
5. perpendicular
6. 75°

S1. rectangle
S2. triangle
1. square, rectangle,
   parallelogram
2. rectangle, parallelogram
3. trapezoid
4. triangle
5. trapezoid
6. square, rectangle,
   parallelogram
7. parallelogram
8. rectangle, parallelogram
9. triangle
10. trapezoid

Problem Solving:
    7 boxes, 6 left over

Page 19

Review Exercises
1. 76°
2. rectangle, square,
   parallelogram
3. 68°
4. answers vary
5. answers vary
6. answers vary

S1. square, rectangle,
    parallelogram
S2. rectangle, parallelogram
1. trapezoid
2. triangle
3. trapezoid
4. trapezoid
5. square, rectangle,
   parallelogram
6. rectangle, parallelogram
7. parallelogram
8. parallelogram
9. triangle
10. trapezoid

Problem Solving: 57 desks

Page 20

Review Exercises
1. triangle
2. trapezoid
3. square, rectangle, trapezoid, parallelogram
4. square, rectangle, parallelogram
5. 72°
6. 44°

S1. scalene, right
S2. equilateral, acute
1. scalene, obtuse
2. isosceles, acute
3. isosceles, right
4. scalene, acute
5. equilateral, acute
6. scalene, obtuse
7. scalene, right
8. isosceles, acute
9. equilateral, acute
10. isosceles, right

Problem Solving: 180°

Page 21

Review Exercises
1. obtuse
2. equilateral
3. scalene
4. isosceles
5. 60°
6. scalene

S1. equilateral, acute
S2. scalene, acute
1. isosceles, right
2. scalene, obtuse
3. equilateral, acute
4. isosceles, acute
5. scalene, right
6. isosceles, right
7. equilateral, acute
8. scalene, acute
9. isosceles, acute
10. scalene, right

Problem Solving: 41°

Page 22

Review Exercises
1. scalene
2. right
3. isosceles, acute
4. 4,536
5. 156
6. 166

S1. 34 ft.
S2. 29 ft.
1. 47 ft.
2. 48 ft.
3. 33 ft.
4. 54 ft.
5. 70 ft.
6. 41 cm.
7. 225 mi.
8. 34 ft.
9. 86 ft.
10. 34 ft.

Problem Solving: 54 ft.

Page 23

Review Exercises
1. obtuse
2. acute
3. right
4. acute
5. scalene
6. 63°

S1. 26 in.
S2. 38 cm.
1. 24 ft.
2. 22 cm.
3. 17 ft.
4. 44 in.
5. 90 cm.
6. 21 ft.
7. 168 cm
8. 27 ft.
9. 32 in.
10. 89 cm

Problem Solving: 102 ft.

Page 24

Review Exercises
1. 42 ft.
2. 76 in.
3. 60 ft.
4. 96 ft.
5. 40°
6. 117°

S1. 100 ft.
S2. 520 ft.
1. 384 ft.
2. 108 in.
3. 39 ft.
4. 171 in.
5. 23 ft.
6. $720
7. 97 in.

Problem Solving: $4.99

Page 25

Review Exercises
1. 128 ft.
2. 104 ft.
3. 85 ft.
4. 339 in.
5. 102 in.
6. 412 ft.

S1. 58 ft.
S2. 24 sections
1. 68 ft.
2. 16 in.
3. 8 hrs.
4. 12 ft.
5. 190 ft.
6. 126 in.
7. 48 in.

Problem Solving:
450 miles per hour

Page 26

Review Exercises
1. scalene
2. right
3. 167°
4. 108 ft.
5. 60°
6. 16 ft.

S1. diameter
S2. $\overline{VT}, \overline{YR}, \overline{XS}$
1. radius
2. chord
3. $\overline{CD}, \overline{DE}, \overline{DG}, \overline{DF}$
4. $\overline{AB}, \overline{GF}, \overline{CE}$
5. 8 ft.
6. P
7. $\overline{RY}, \overline{VT}, \overline{XS}$
8. 48 ft.
9. $\overline{PX}, \overline{PS}, \overline{PZ}$
10. $\overline{XS}$

Problem Solving: 13 mi.

Page 27

Review Exercises
1. 32 ft.
2. · 9 ft.
3. 18.84
4. 66
5. 50°
6. scalene, right

S1. radius
S2. $\overline{AH}, \overline{AB}, \overline{AE}$
1. chord
2. chord
3. $\overline{GF}, \overline{HE}, \overline{CD}$
4. $\overline{XT}$
5. 36 ft.
6. 48 in.
7. $\overline{WX}, \overline{WT}, \overline{WQ}$
8. $\overline{RS}, \overline{XT}$
9. A
10. $\overline{AH}, \overline{AE}$

Problem Solving: $672

Solutions

Page 28

Review Exercises
1. 24 ft.
2. 110
3. 25.12
4. 28 in.
5. scalene
6. isosceles

S1. 12.56 ft.
S2. 44 ft.
1. 18.84 ft.
2. 25.12 ft.
3. 28.26 ft.
4. 88 ft.
5. 37.68 ft.
6. 31.4 ft.
7. 12.56 ft

Problem Solving: $5,600

Page 29

Review Exercises
1. 292 in.
2. 100 ft.
3. 315 m.
4. 87 ft.
5. 368 ft.
6. 672 ft.

S1. 88 ft.
S2. 28.26 ft.
1. 47.1 ft.
2. 50.24 ft.
3. 66 ft.
4. 132 ft.
5. 110 ft.
6. 37.68 ft.
7. 314 ft.

Problem Solving: 88 ft.

Page 30

Review Exercises
1. 88 ft.
2. 90 ft.
3. 44 ft.
4. 8
5. 21
6. obtuse

S1. 110 ft.
S2. 44 in.
1. 44 in.
2. 88 mi.
3. 22 ft.
4. 4 ft.
5. 3 ft.
6. 22 ft.
7. 31.4 ft.

Problem Solving: 124 ft.

Page 31

Review Exercises
1. 67°
2. 83°
3. trapezoid
4. 28
5. 12
6. 71°

S1. 35 ft.
S2. 15 ft.
1. 28 ft.
2. 352 m. or 351.68 m.
3. 25,120 mi.
4. 31.4 ft.
5. 176 ft.
6. 84 ft.
7. 20 ft.

Problem Solving: 240 ft.

Page 32

Review Exercises
1. 170 ft.
2. 18.84 ft.
3. 192 ft.
4. 88 ft.
5. 51 ft.
6. 76 ft.

S1. 54 ft.
S2. 88 ft.
1. 66 ft.
2. 28 ft.
3. 116 ft.
4. 21 ft.
5. 51 ft.
6. 25.12 ft.
7. 177 ft.
8. 48 ft.
9. 460 ft.
10. 40 ft.

Problem Solving: 7 hrs.

Page 33

Review Exercises
1. acute
2. obtuse
3. right
4. 8
5. pentagon
6. decagon

S1. 116 ft.
S2. 37.68
1. 110 ft.
2. 258 ft.
3. 88 ft.
4. 185 ft.
5. 460 ft.
6. 390 ft.
7. 284 ft.
8. 132 ft.
9. 252 ft.
10. 132 ft.

Problem Solving: $306

Page 34

Review Exercises
1. answers vary
2. answers vary
3. 75°
4. 165°
5. 45°
6. 127°

S1. 110 ft.
S2. 85 ft.
1. 210 ft.
2. 49 ft.
3. 720 ft.
4. 38 ft.
5. $1,400
6. 156 ft.
7. 184 ft.

Problem Solving:
29,050 gallons

Page 35

Review Exercises
1. 64 ft.
2. 90 ft.
3. 18.84 ft.
4. 66 ft.
5. 140 ft.
6. 99 ft.

S1. 63 ft.
S2. 3 ft.
1. 125 laps
2. 4,605 ft.
3. 192 in.
4. 12 ft.
5. 4 laps
6. 304 ft.
7. 66 ft.

Problem Solving: 160 mi.

Page 36

Review Exercises
1. answers vary
2. acute, obtuse, right, straight
3. radius, diameter, chord, center
4. scalene, isosceles, equilateral
5. acute, obtuse, right
6. pentagon, hexagon, octagon, decagon

S1. 5 ft.
S2. 9 cm.
1. 7 ft.
2. 10 ft.
3. 8 ft.
4. 19 ft.
5. 9 ft.
6. 12 ft.
7. 16 ft.

Problem Solving: $14.15

Page 37

Review Exercises
1. 68 ft.
2. 52 ft.
3. 174 ft.
4. 37.68 ft.
5. 22 ft.
6. 525 ft.

S1. 8 ft.
S2. 13 ft.
1. 9 ft.
2. 8 ft.
3. 18 cm.
4. 28 cm.
5. 13 ft.
6. 35 cm.
7. 20 ft.

Problem Solving: 230 cartons

Page 38

Review Exercises
1. 68 ft.
2. 476 ft.
3. 132 ft.
4. 25.12 ft.
5. 6
6. 8

S1. 8 ft.^{2}
S2. 19 ft.^{2}
1. 16 ft.^{2}
2. 18 ft.^{2}
3. 27 ft.^{2}
4. 21 ft.^{2}
5. 20 ft.^{2}
6. 24 ft.^{2}
7. 13 ft.^{2}

Problem Solving: 92

Page 39

Review Exercises
1. equilateral
2. obtuse
3. 9 ft.
4. 28.26 ft.
5. 132 ft.
6. 288 ft.

S1. answers vary
S2. answers vary
1. answers vary
2. answers vary
3. answers vary
4. answers vary
5. answers vary
6. answers vary
7. answers vary

Problem Solving: $21,200

Page 40

Review Exercises
1. 84°
2. 163°
3. square, rectangle, parallelogram
4. 145 ft.
5. 64 ft.
6. 12.56 ft.

S1. 169 sq. ft.
S2. 165 sq. ft.
1. 84 sq. ft.
2. 400 sq. ft.
3. 132 sq. ft.
4. 391 sq. ft.
5. 13.5 sq. ft.
6. 625 sq. ft.
7. 32,400 sq. ft.

Problem Solving: 842

Page 41

Review Exercises
1. 143 sq. ft.
2. 225 sq. ft.
3. 62 ft.
4. 36 ft.
5. 88 ft.
6. 31.4 ft.

S1. 209 ft.^{2}
S2. 256 ft.^{2}
1. 126 ft.^{2}
2. 625 ft.^{2}
3. 475 ft.^{2}
4. 400 ft.^{2}
5. 14 ft.
6. 25 ft.
7. 30,000 cm.^{2}

Problem Solving: 285 sq. ft.

Page 42

Review Exercises
1. 68 ft.
2. 289 ft.^{2}
3. 160 ft.^{2}
4. 52 ft.
5. 11 ft.
6. 50°

S1. 78 ft.^{2}
S2. 77 ft.^{2}
1. 54 ft.^{2}
2. 176 ft.^{2}
3. 17.5 ft.^{2}
4. 91 ft.^{2}
5. 84 ft.^{2}
6. 117 ft.^{2}
7. 71.5 ft.^{2}

Problem Solving: $66,000

Page 43

Review Exercises
1. 60 ft.
2. 76 ft.
3. 66 ft.^{2}
4. 42 ft.^{2}
5. 126 ft.^{2}
6. 37.68 ft.

S1. 17.5 ft.^{2}
S2. 65 ft.^{2}
1. 64 ft.^{2}
2. 325 ft.^{2}
3. 71.5 ft.^{2}
4. 264 ft.^{2}
5. 36 ft.^{2}
6. 120 cm.^{2}
7. 16 ft.

Problem Solving: 320 ft.^{2}

Solutions

Page 44

Review Exercises
1. 196 ft.²
2. 96 ft.²
3. 54 ft.²
4. 105 ft.²
5. 52.5 ft.²
6. 64 ft.²

S1. 25 ft.²
S2. 40 ft.²
1. 36 ft.²
2. 44 ft.²
3. 70 ft.²
4. 144 ft.²
5. 80 ft.²
6. 16.5 ft.²
7. 330 ft.²

Problem Solving: 240 in.²

Page 45

Review Exercises
1. 32 ft.
2. 46 ft.
3. 85 ft.
4. 88 ft.
5. 230 ft.
6. 132 ft.

S1. 49 ft.²
S2. 30 ft.²
1. 85 ft.²
2. 800 cm.²
3. 1,650 ft.²
4. 64 ft.²
5. 26 ft.²
6. 9 ft.²
7. 42 cm.²

Problem Solving:
2,505 pounds

Page 46

Review Exercises
1. 54 ft.²
2. 66 cm.²
3. 38 ft.
4. 56 ft.
5. 196 ft.²
6. 84 ft.²

S1. 50.24 ft.²
S2. 113.04 ft.²
1. 78.5 ft.²
2. 616 ft.²
3. 12.56 ft.²
4. 200.96 ft.²
5. 78.5 ft.²
6. 113.04 ft.²
7. 154 ft.²

Problem Solving: 320 ft.

Page 47

Review Exercises
1. obtuse
2. acute
3. 110°
4. equilateral, acute
5. 18.84 ft.
6. 28.26 ft.²

S1. 1,386 ft.²
S2. 28.26 ft.²
1. 314 ft.²
2. 1,256 ft.²
3. 616 ft.²
4. 3.14 ft.²
5. 254.34 ft.²
6. 1,386 ft.²
7. 379.94 ft.²

Problem Solving: 20 classes

Page 48

Review Exercises
1. 154 ft.²
2. 196 ft.²
3. 135 ft.²
4. 31.5 ft.²
5. 96 ft.²
6. 72 ft.²

S1. P = 38 ft., A = 84 ft.²
S2. P = 27 ft., A = 25 ft.²
1. P = 48 ft., A = 144 ft.²
2. P = 44 ft., A = 120 ft.²
3. P = 38 ft., A = 72 ft.²
4. C = 18.84 ft.,
   A = 28.26 ft.²
5. C = 44 ft., A = 154 ft.²
6. P = 24 ft., A = 24 ft.²
7. P = 27 ft., A = 37.5 ft.²

Problem Solving: $396

Page 49

Review Exercises
1. 60 ft.
2. 84 ft.
3. 25.12 ft.
4. 36 ft.
5. 44 ft.
6. 51 ft.

S1. P = 23 ft., A = 30 ft.²
S2. C = 88 ft., A = 616 ft.²
1. P = 32 ft., A = 54 ft.²
2. P = 27 ft., A = 35 ft.²
3. P = 48 ft., A = 144 ft.²
4. C = 25.12 ft.,
   A = 50.24 ft.²
5. P = 86 ft., A = 406 ft.²
6. P = 47 ft., A = 112.5 ft.²
7. P = 28 ft., A = 42 ft.²

Problem Solving: $1.35

Page 50

Review Exercises
1. 256 ft.²
2. 72 ft.²
3. 352 ft.²
4. 126 ft.²
5. 288 ft.²
6. 78.5 ft.²

S1. 52 ft.
S2. 20 bags
1. 96 ft.
2. 14 sections
3. 50.24 ft.
4. 3 cans
5. 156 ft.
6. 16 mi.
7. 630 meters

Problem Solving: $1,200

Page 51

Review Exercises
1. 55°
2. 102°
3. 36 ft.
4. 235 ft.
5. 24 ft.
6. P = 138 ft., A = 884 ft.²

S1. 6.28 mi.
S2. $1,440
1. 132 mi.
2. 40,000 sq. ft.
3. 96 in.
4. 28.26 in.²
5. 144 in.
6. 56 ft.²
7. 144 in.²

Problem Solving: 110 minutes

Solutions

Page 52

Review Exercises
1. 25.12 ft.
2. 112 ft.²
3. 380 cm.
4. 48 ft.²
5. 66 ft.
6. 82.5 ft.²

S1. rectangular prism,
   6, 12, 8
S2. square pyramid,
   5, 8, 5
1. cone, 1, 1, 1
2. cube, 6, 12, 8
3. cylinder, 2, 2, 0
4. triangular pyramid,
   4, 6, 4
5. triangular prism,
   5, 8, 6
6. sphere
7. 1

Problem Solving: $7.68

Page 53

Review Exercises
1. square pyramid,
   triangular prism
2. cube, rectangular prism
3. sphere
4. cube, rectangular prism
5. square pyramid
6. cone, cylinder

S1. answers vary
S2. answers vary
1. answers vary
2. answers vary
3. answers vary
4. answers vary
5. answers vary
6. answers vary
7. answers vary
8. answers vary
9. answers vary
10. answers vary

Problem Solving: 180 in.

Page 54

Review Exercises
1. 72°
2. 66°
3. 78.5 ft.²
4. 31.5 ft.²
5. 90 ft.
6. 256 ft.²

S1. 24 ft.²
S2. 136 ft.²
1. 150 m.²
2. 176 ft.²
3. 294 ft.²
4. 356 cm.²
5. 258 cm.²
6. 30 m.²
7. 280 in.²

Problem Solving: 84 mi.²

Page 55

Review Exercises
1. 50.24 ft.²
2. 28.26 ft.²
3. 441 ft.²
4. 540 cm.²
5. 330 ft.²
6. 104 ft.²

S1. 34 ft.²
S2. 96 m.²
1. 40 ft.²
2. 216 ft.²
3. 270 m.²
4. 66 ft.²
5. 90 in.²
6. 166 cm.²
7. 62 in.²

Problem Solving: $19,200

Page 56

Review Exercises
1. 116 ft.
2. 46 ft.
3. 44 ft.
4. 18.84 ft.
5. 70 ft.
6. 114 ft.

S1. 336 cm.³
S2. 64 ft.³
1. 240 cm.³
2. 343 in.³
3. 4,480 in.³
4. 80 ft.³
5. 273 cm.³
6. 1,680 ft.³
7. 2,808 ft.³

Problem Solving: 4,550 mi.

Page 57

Review Exercises
1. 142 ft.²
2. 192 ft.³
3. 52°
4. 31°
5. 27°
6. 31.4 cm.

S1. 360 in.³
S2. 343 ft.³
1. 560 ft.³
2. 729 in.³
3. 1,200 ft.³
4. 1,331 cm.³
5. 4,096 in.³
6. 2,200 in.³
7. 280 ft.³

Problem Solving: $117,000

Page 58

Review Exercises
1. 242 ft.²
2. 84 in.³
3. 96 ft.²
4. 64 ft.³
5. 228 cm.
6. 361 ft.²

S1. 125 m.³
S2. 150 m.²
1. 258 cm.²
2. 270 cm.³
3. 8 ft.³
4. 24 ft.²
5. 126 ft.²
6. 90 ft.³
7. 216 ft.³

Problem Solving: 6 days

Page 59

Review Exercises
1. 72 ft.²
2. 342 cm.²
3. 256 in.²
4. 135 in.²
5. 40.5 ft.²
6. 50.24 ft.²

S1. 756 ft.³
S2. 96 ft.²
1. 412 ft.²
2. 486 in.²
3. 1,500 in.³
4. 148 in.²
5. 13 in.

Problem Solving: $33,120

Page 60	Page 61	Page 62	Page 63
Review Exercises	Review Exercises	Review Exercises	Review Exercises
1. answers vary	1. answers vary	1. 52 ft.	1. 53 ft.
2. answers vary	2. answers vary	2. 384 ft.	2. 88 ft.
3. answers vary	3. answers vary	3. 44 ft.	3. 166 ft.
4. answers vary	4. answers vary	4. 174 ft.	4. 34.54
5. answers vary	5. △BCD, △FGH, △ABF	5. 114 ft.	5. 260 ft.
6. triangle FIK	6. trapezoid GHJI	6. 37.68 ft.	6. 290 ft.
7. answers vary	7. answers vary	7. 240 ft.^{2}	7. 32.5 ft.^{2}
8. answers vary	8. pentagon BDEHF	8. 56 ft.^{2}	8. 154 ft.^{2}
9. answers vary	9. answers vary	9. 192 ft.^{2}	9. 28.26 ft.^{2}
10. trapezoid LKIJ	10. answers vary	10. 169 ft.^{2}	10. 119 ft.^{2}
11. equilateral, acute	11. scalene, acute	11. 45 ft.^{2}	11. 136 ft.^{2}
12. scalene, obtuse	12. scalene, right	12. 154 ft.^{2}	12. 27.5 ft.^{2}
13. $\overline{CG}, \overline{AF}$	13. hypotenuse	13. 50.24 ft.^{2}	13. 27 ft.^{2}
14. $\overline{EC}, \overline{EF}, \overline{EA}, \overline{EG}$	14. trapezoid	14. 49.5 ft.^{2}	14. 256 ft.^{2}
15. $\overline{BH}, \overline{CG}, \overline{AF}$	15. square, rectangle, parallelogram	15. 48 ft.^{2}	15. 17.5 ft^2
16. 12 ft.	16. octagon	16. 289 ft.^{2}	16. 25 ft.^{2}
17. chord	17. regular	17. 216 cm.^{2}	17. 100 ft.^{2}
18. rectangular prism, 6, 12, 8	18. square pyramid, 5, 8, 5	18. 152 ft.	18. 720 ft.^{2}
19. triangular prism, 5, 9, 6	19. cylinder 2, 2, 0	19. 236 cm.^{2}	19. 112 ft.^{2}
20. cube, 6, 12, 8	20. cone, 1, 1, 1	20. 750 cm.^{3}	20. 216 ft.^{3}

Glossary

A

Acute Angle - Any angle whose measure is between 0° and 90°.

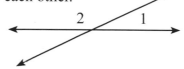

Acute Triangle - A triangle with each angle measuring less than 90°.

Adjacent Angles - Angles that are next to each other.

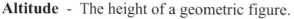

∠1 and ∠2 are adjacent angles.

Altitude - The height of a geometric figure.

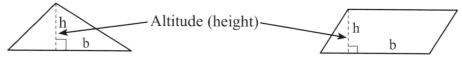

Angle - The figure formed when two rays share a common endpoint.

Area - The number of square units that a region contains.

B

Base - The side of a geometric figure that contains one end of the altitude (height).

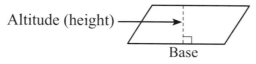

C

Chord - A line segment that connects any two points of a circle.

Circle - A closed curve with all its points in one place and the same distance from a point that is called the center.

Circumference - The distance around a circle. The perimeter of a circle.

Collinear Points - Points that lie on the same line.

Complementary Angles - Two angles whose measures have a sum of 90°.

Cone - A 3-dimensional figure whose base is a circle joined to a vertex by a curved surface.

Glossary

Congruent - Having exactly the same shape and size.

Coplanar - Located in the same plane.

Cube - A solid figure with six faces. Every face is a square and every edge is one of the same length.

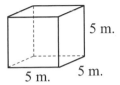

Cubic Unit - A unit of measure that has length, width, and depth, and is used to measure volume. Examples are cubic feet and cubic meters.

Cylinder - A solid figure consisting of two congruent parallel circles joined by a curved surface.

D

Decagon - A ten-sided polygon.

Degree - A unit used to measure angles.

Diagonal - A line segment that joins two vertices of a polygon.

Line segment $\overline{BC}$ is a diagonal.

Diameter - A line segment that passes through the center of a circle connecting two points on the circle.

Line segment $\overline{AB}$ is a diameter.

E

Edge - A line segment where two faces of a solid figure meet.

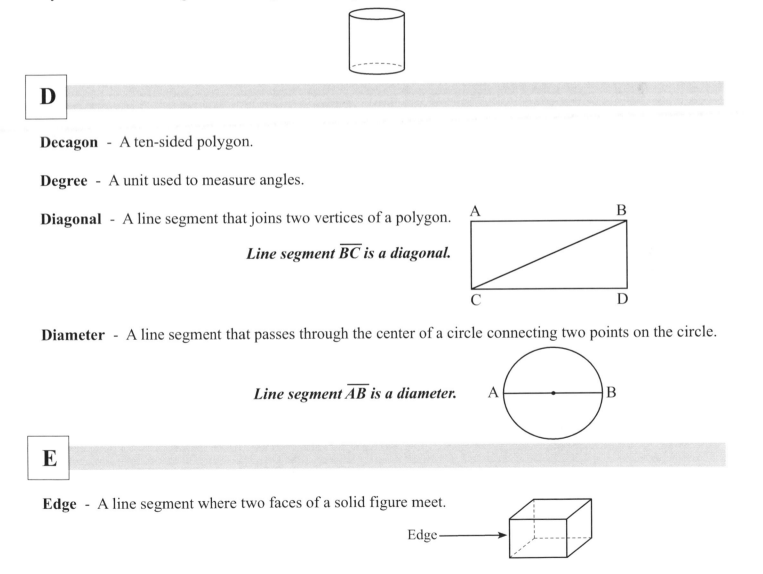

Glossary

Endpoint - The point at the end of a ray or line segment.

Equidistant - The same distance from something.

Equilateral - A figure in which all sides have the same length. All sides are congruent.

Equilateral triangle - A triangle with three congruent sides. All three sides are the same length.

F

Face - A plane figure that serves as one side of a solid figure.

Face →

Figure - A closed geometric shape with two or three dimensions.

Plane Figure   Solid Figure

G

Geometry - The branch of mathematics that deals with points, lines, angles, surfaces, and solid shapes.

H

Height - The altitude of a geometric figure.

Hexagon - A six-sided polygon.

Hypotenuse - The side opposite the right angle in a right triangle. The longest side of a right triangle.

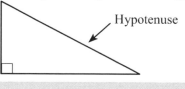

Hypotenuse

I

Intersect - When figures meet or cross.

Irregular polygon - A polygon whose sides are not all the same length.

Isosceles Triangle - A triangle with at least two congruent sides.

Glossary

Leg of a right triangle - A side of a right triangle that forms the right angle.

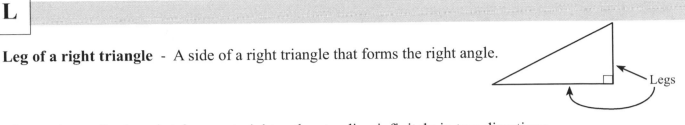

Legs

Line - A set of points that form a straight path extending infinitely in two directions.

Line of symmetry - A line dividing a figure into two congruent parts that are mirror images of each other.

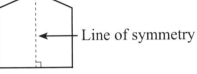

Line of symmetry

Line segment - A part of a line. A line segment has two endpoints.

M

Midpoint - The point on a line segment that divides it into two congruent line segments.

N

Nonagon - A nine-sided polygon.

O

Obtuse angle - An angle whose measure is greater than 90° but less than 180°.

120°

Octagon - An eight-sided polygon.

Opposite angles - Angles in a quadrilateral that do not have a common side.

∠**B** and ∠**C** are opposite angles.

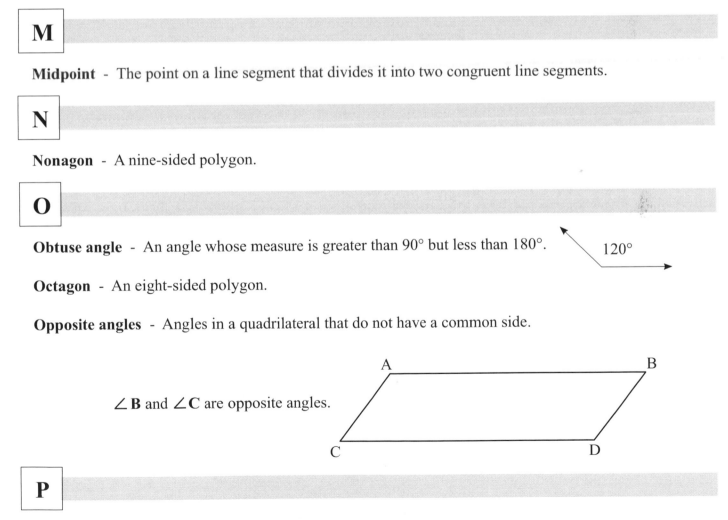

P

Parallel - Always the same distance apart.

Glossary

Parallel lines - Lines lying in the same plane that stay the same distance apart.

Parallelogram - A quadrilateral with both pairs of opposite sides parallel and congruent.

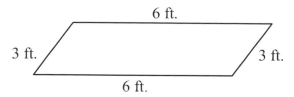

Pentagon - A five-sided polygon.

Perimeter - The distance around the figure. The sum of the lengths of the sides of a polygon.

Perpendicular - Forming right angles. Two line segments, rays, or lines that intersect at a right angle are perpendicular.

Pi - The ratio of the circumference of a circle to its diameter. Written π. The value of is approximately equal to 3.14 as a decimal, and $\frac{22}{7}$ as a fraction.

Plane - A flat surface extending infinitely in all directions.

Plane Figure - A two-dimensional figure that lies entirely in one plane. Examples are circles, polygons, and angles.

Point - An exact position in space often represented by a dot.

Polygon - A closed figure composed of line segments that meet at their endpoints.

Prism - A three-dimensional figure with one pair of opposite faces that are parallel and of congruent, both which are polygons. The rest of the faces are parallelograms.

Pyramid - A three-dimensional figure whose base is a polygon. The other faces are triangles that meet at a common vertex.

Pythagorean Theorem - In a right triangle, the sum of the squares of the lengths of the two legs is equal to the square of the length of the hypotenuse.

$$a^2 + b^2 = c^2$$

Glossary

Q

Quadrilateral - A four-sided polygon.

R

Radius - The line segment that extends from the center of a circle to any point on the circle.

Line segment $\overline{AB}$ is a radius.

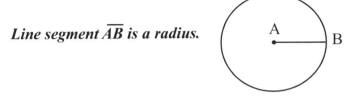

Ray - A part of a line that has one endpoint and extends infinitely in one direction.

Rectangle - A quadrilateral with two pairs of parallel, congruent sides, and four right angles.

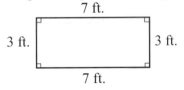

Rectangular Prism - A prism with six faces, all of which are rectangles.

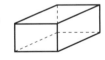

Regular polygon - A polygon that has all congruent sides and all congruent angles.

Rhombus - A parallelogram with all four sides congruent.

Right angle - An angle whose measure is 90°.

Right triangle - A triangle that has one right angle.

S

Scalene triangle - A triangle with no sides congruent.

Sides - The line segments that form a polygon.

Similar - Figures that have the same shape, but not necessarily the same size.

Glossary

Solid figure - A geometric, three-dimensional figure. Examples are prisms, spheres, and pyramids.

Sphere - A three-dimensional figure formed by all the points that are the same distance from a fixed point called the center.

Square - A parallelogram with four congruent sides and four right angles.

Square unit - A unit of measure such as a square inch or square meter, used to measure area.

Straight angle - An angle whose measure is 180°.

Supplementary Angles - Two angles whose measures have a sum of 180°.

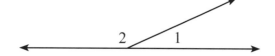

∠1 and ∠ 2 are supplementary because their sum is 180°.

Surface area - The total area of the surface of a solid figure.

T

Three-dimensional - Relating to geometric figures with length, width, and depth. Cones, spheres, and prisms are examples of figures with three dimensions.

Trapezoid - A quadrilateral with exactly one pair of parallel sides.

Triangle - A three-sided polygon.

Two-dimensional - Relating to geometric figures that have length and width, but not depth. Polygons and circles are examples of figures with two-dimensions.

V

Vertex - A point at which two lines, rays, or line segments meet to form an angle.

Vertical angles - Opposite angles formed when two lines intersect. Vertical angles are congruent.

∠1 and ∠3 are vertical angles. So are ∠2 and ∠4.

Volume - The number of cubic units it takes to fill a solid figure.

78